I0868346

The Army Surveys *of* Gold Rush California

The Army Surveys *of* Gold Rush California

Reports of the Topographical Engineers 1849–1851

Edited by
Gary Clayton Anderson and
Laura Lee Anderson

THE ARTHUR H. CLARK COMPANY
An imprint of the University of Oklahoma Press
Norman, Oklahoma
2015

LIBRARY OF CONGRESS CATALOGING-IN-PUBLICATION DATA
The Army surveys of Gold Rush California : reports of topographical engineers, 1849–1851 / edited by Gary Clayton Anderson and Laura Lee Anderson.
pages cm
Includes bibliographical references and index.
ISBN 978-0-87062-430-8 (hardback)
1. California—Gold discoveries—Sources. 2. California—History—1846–1850—Sources. 3. California—Environmental conditions—History—19th century—Sources. 4. Frontier and pioneer life—California—Sources. 5. Topographical surveying—California—History—19th century—Sources. 6. California—Geography—Sources. 7. Derby, George Horatio, 1823–1861—Diaries. 8. Engineers—California—Diaries. 9. United States. Army—Officers—Diaries. I. Derby, George Horatio, 1823–1861. II. Anderson, Gary Clayton, 1948– III. Anderson, Laura Lee, 1950–
F865.A76 2015
979.4'04—dc23

2014041737

The paper in this book meets the guidelines for permanence and durability of the Committee on Production Guidelines for Book Longevity of the Council on Library Resources, Inc. ∞

Copyright © 2015 by the University of Oklahoma Press, Norman, Publishing Division of the University. Manufactured in the U.S.A.

All rights reserved. No part of this publication may be reproduced, stored in a retrieval system, or transmitted, in any form or by any means, electronic, mechanical, photocopying, recording, or otherwise—except as permitted under Section 107 or 108 of the United States Copyright Act—without the prior written permission of the University of Oklahoma Press.

To request permission to reproduce selections from this book, write to Permissions, University of Oklahoma Press, 2800 Venture Drive, Norman OK 73069, or email rights.oupress@ou.edu.

1 2 3 4 5 6 7 8 9 10

to

William Swagerty

bon vivant,

fellow graduate of the

Newberry Library "school" of Indian history,

a fine California historian,

and our friend

Contents

Maps

Complete, full-size versions of the maps named above, along with two others—Frémont's Map of Oregon and Upper California, 1848 *and* The Colorado River, 1850—*may be found folded and inserted at the back of the casebound edition of this volume.*

Acknowledgments

Some years ago, while working on the second floor of the National Archives, Laura and I encountered a number of fascinating journals. At the time, we were working to improve a chapter on early California for an upcoming book and had ordered up the Letters Received file from Record Group 393, Department of the Pacific. This collection, massive in size (more than two hundred boxes), includes the early military correspondence of the U.S. Army in California, starting in 1848. It soon became obvious that the documents provide significant, little-known firsthand observations of gold rush–era California; in particular, they offer excellent descriptions of the American Indian peoples of the regions north and east of Los Angeles. Most of the journals were kept by topographical engineers, but on occasion, a regular line officer would do so. Some run to more than fifty pages, while others to only fifteen or twenty. We photocopied a couple of the journals, and were impressed upon reading them with how well they were written and with the serious discussions they contain.

When we returned to Norman, Bob Clark of the Arthur H. Clark Company asked to see one. Written by George Horatio Derby, this journal had illustrations that included a massive map of the Tulares Region of southern California. Bob was dumbfounded, asking, "Do you know who Derby is?" When Laura and I looked a bit crestfallen, Bob described Derby's career as a humorist in early California and the early books written about him. I then dropped the bombshell: there were three other massive Derby journals in the archives, all written between 1849 and 1851. And other journals, written by both engineers and officers, complemented Derby's work. We had counted ten that were significant. Sitting in Bob's office, we then agreed to gather and publish them, and that is how this book came together.

As the project took shape, it became clear that little was known about Derby's early career. One secondary source placed him in California at about 1853, when his literary career began. Just two of the four army journals of his included in this volume have ever been published, and only in obscure places, and six journals by others are published here for the first time. Given the few sources available, Laura and I turned to friends for assistance. Peter Blodgett, a specialist in early California and curator at the Henry E. Huntington Library was of immense help in putting together references that aided the editing. Original copies of the spectacular maps that Derby and the other engineers made (all reproduced here at the back of the clothbound book) turned up at various libraries and archives, including the Library of Congress, the Amon Carter Museum, the Brigham Young University Library, and the Huntington. Many curators helped obtain digital copies for reproduction. Emmy Ezzell and Anna María Rodríguez at the University of Oklahoma Press spent hours working to make the maps serviceable, one of which had to be completely redrawn by cartographer Tom Jonas. Given the substantial cost of such reproductions, it was necessary to ask for financial assistance. At the University of Oklahoma, the Vice President for Research, the Provost, the Dean of Arts and Sciences, and the Chair of the History Department all helped to make publication of the maps possible. We thank them profusely.

The project itself, begun by Bob Clark, was brought to final completion by Editor-in-Chief Chuck Rankin, a friend and colleague for many years. Chuck helped sort out all sorts of technical problems that appear when doing a book that is part history and part geography. For an expert reading of the manuscript, we turned to another old friend, William R. Swaggerty, of the University of the Pacific. An editor and historian of considerable note, as well as an expert on early

California history, Bill threw his heart into the project, identifying obscure sources that explained both major and minor details. Knowing Bill since our wonderful year together at the Newberry Library in 1980, Laura and I agreed that we would dedicate the book to him. He earned it. Given the fact that so many California historians have passed before us without access to these valuable resources, we hope that the journals will provide a better understanding of early California—to Euro-Americans a place of some mystery in 1849, when the first topographical engineer hit the trail to explore a land with so many daunting landscapes and so many exotic people.

Gary Clayton and Laura Lee Anderson
University of Oklahoma

Introduction: The Army Journalists

As the guns of war in Mexico fell silent in 1848, the United States government slowly turned to administering the vast new empire it had acquired in the West. California received special attention after millworkers discovered gold in the American River that summer. Although so-called volunteers, including the famous Mormon Battalion, were already in the new region at the close of the war, regular army troops arrived in 1848 and 1849 at Monterey, which had been the headquarters for the military government. The troops assigned to California came from the 2nd U.S. Infantry and elements of the 3rd Artillery. Within this command were a host of officers, quite a number of whom had been in the army for more than a decade. Most would later serve in the Civil War, and a few, such as the young lieutenant William Tecumseh Sherman, would become quite famous. But in California in 1849, most found promotion to be a very slow process. A few had been advanced to first lieutenant, while most remained second lieutenants and some were experiencing duty on the "frontier" for the first time.

When the troops arrived in California, to most of these officers, the new lands seemed different from any they had known before. Franciscan missionaries from Mexico had reached the area by 1769. They built a string of more than a dozen missions along the coast, reaching northward all the way into the Sonoma valley above San Francisco. While the padres had some success in bringing Indians into the missions—or using military troops to concentrate them nearby—they had been forced to relinquish mission lands in 1833–34, when the Mexican government secularized the missions. A small Hispanic ranching community then replaced the missions, with large land grants being given out between 1834 and 1846. Nevertheless, for the most part, the ranchers had very little impact on most of California. The immense Central Valley (the San Joaquin and Sacramento river valleys) and the Sierra Nevada to the east, as well as the regions north of Sonoma, were Indian country. Several hundred Native bands inhabited these lands, and much of this region remained unexplored by either Mexicans or Anglo-Americans.

A few Americans had reached California before the regular army, creating some of the first maps of the region. John C. Frémont had gained a reputation as a pathfinder by leading several expeditions into the mountains and beyond. Frémont, a topographical engineer by training, mapped much of the Oregon Trail in 1845, before turning south and entering California, where he was most famous for leading the so-called Bear Flag Revolt the following year. Frémont had been accompanied by professional mapmaker Charles Preuss since St. Louis, and he brought Preuss along with him to California. The map that resulted from the trip (reproduced here) later became a standard guide for emigrants who headed to the new "golden" lands of California via the Oregon and California Trails.[1] Yet another survey had occurred, far to the south along the New Mexico–Arizona-California road that led to San Diego. Topographical engineer Lieutenant William H. Emory had joined General Stephen W. Kearny in his march through the southwest in 1846, penning a journal and making a map of the road that thousands of future emigrants used to reach the Pacific Slope.[2]

---

[1]See John Charles Frémont, *Geographical Memoir upon Upper California, in Illustration of His Map of Oregon and California, Addressed to the Senate of the United States* (Washington, D.C.: Wendell and Van Benthuysen, 1848), also printed in Senate Miscellaneous Document 148, 30th Congress, 1st session. Frémont's maps, including the one of Upper California, were included in Joseph Ware, *The Emigrants' Guide to California,* ed. John Caughey (1849; reprint, Princeton: Princeton University Press, 1932).

[2]See W. H. Emory, *Notes of a Military Reconnoissance from Fort Leavenworth, in Missouri, to San Diego, in California, Including Parts of Arkansas, Del Norte, and Gila Rivers* (Washington, D.C.: Wendell and Van Benthuysen, 1848).

As the new commanding officer of troops in California, General Persifor F. Smith, realized, these efforts hardly offered much information. As the threat of Mexican invasion subsided in spring 1849, Smith wanted to know the geography of California, places that men like Frémont had not penetrated. He needed to know where military forts might be needed, and this included some understanding of roads and bridges that would be necessary to supply them, especially garrisons in the mountain country north of Sonoma and east of the Central Valley. Smith quickly realized that the mission of the military in California had changed from defending the region from international threat to keeping the peace domestically by protecting the mining frontier and separating Indians and miners from each other in order to prevent war. Indeed, by fall it also became obvious that some Indian groups, decimated by disease, needed protection from miners, who were overrunning river valleys and destroying the salmon fishing habitat.[3]

Former fur trappers and early pioneers who had settled in various coastal regions, including such towns as Los Angeles and Monterey, had entered some of the lands controlled by Natives. What they could say of these lands and the Indians who lived in them was helpful but supplied little information to the army. One of the most famous, William Wolfskill, who had arrived in California in 1831, was more interested in making wine than guiding military expeditions.[4] These men often overestimated the numbers of Indians, and they seemed unaware that disease—mostly viruses such as measles, smallpox, and malaria, introduced in the 1830s and continuing unabated into the late

[3]Smith's many letters regarding policy in early California are in National Archives Record Group (hereafter NARG) 393, Letters Received (hereafter LR), Department and Division of the Pacific (hereafter DP). His report of October 7, 1849, some thirty-three pages in length, was published in Senate Executive Document 47, 31st Congress, 1st session, part 2, and was also included in Philip T. Tyson, *Geology and Industrial Resources of California* (Baltimore: Wm. Minifie, 1851), 75–108. See also Secretary of War William Crawford to General Smith, April 3; Lieutenant Sherman to Governor R. B. Mason, April 10; Brevet Brigadier General Bennett Riley to General R. Jones, April 25; and Lieutenant Sherman to General Riley, May 4, 1849, NARG 393, LR, DP. General Smith was born in Philadelphia in 1798, taking a degree at Princeton University just eighteen years later. He migrated to New Orleans, fought under General Edmund P. Gaines in the Florida War, and ultimately commanded a division in the Mexican War. General Winfield Scott described Smith as "cool, unembarrassed, and ready." See his biography in James Grant Wilson and John Fiske, eds., *Appletons' Cyclopaedia of American Biography* (New York: D. Appleton, 1888), 5: 583.

[4]See Iris Wilson Engstrand, *William Wolfskill, 1798–1866* (Glendale, Calif.: Arthur H. Clark, 1965).

1840s—was taking a heavy toll on bands. Once settlers began arriving in numbers, newspapers quickly reported almost any incident that occurred with Natives and often offered population estimates, which, once again, frequently exaggerated their numbers.[5]

While historians and anthropologists have occasionally referenced the reports of such newspapers and early pioneers in writing about California Indians, the most frequently cited attempt to marshal these sources and arrive at a demographic assessment of the entire state at the time of the Gold Rush was made by Sherburne F. Cook. After collecting such sources, he argued that 150,000 Indians lived in California in 1845, that these numbers plummeted rapidly to 100,000 by 1850, and fell to just 50,000 by 1855.[6] Most historians and anthropologists have adopted Cook's numbers without questioning them, despite the fact that many of his sources came from editors who had never been in the interior.

General Smith wanted to know exact populations in 1849. He soon set in motion a series of surveys both by topographical engineers, when available, and by regular army officers, who set out to map the entire region and give detailed assessments of Native populations. He also was pressed by the War Department to assess the value of the mineral assets in the region, given the fact that so many early reports seemed exaggerations. Finally, the War Department wanted to know

[5]The newspaper reports were used extensively by anthropologist Sherburne Cook in his demographic work. The two most frequently cited papers used by Cook were the *San Francisco Bulletin* and the *Sacramento Union*. See Cook's detailed assessment in *The Conflict between the California Indian and White Civilization* (Berkeley: University of California Press, 1943), 1–55.

[6]Cook based his numbers on what he called "interpolation," which fundamentally was guesswork based upon extremely unreliable sources. Besides the newspaper accounts, Cook used early memoirs, such as E. Bryant, *What I Saw in California* (1846; reprint, Minneapolis: Ross and Haines, 1967), and J. Carr, *Pioneer Days in California* (Eureka, 1891; reprint, Fresno, Calif.: Mid-Cal Books, 1982). While these can be valuable sources, Cook admits in his study that he was only able to see the "reports" of the commissioner of Indian Affairs and never was able to investigate correspondence in the files of the "California Superintendency," which are extensive. See *The Conflict between the California Indian and White Civilization*, book 2, and Cook, *The Population of the California Indians, 1769–1970* (Berkeley: University of California Press, 1976), 44–45. Yet more speculation on Native populations in California is found in *California*, ed. Robert Heizer, vol. 8 of *Handbook of North American Indians*, gen. ed. William C. Sturtevant, (Washington, D.C.: Smithsonian Institution, 1978–). Cook contributed the article "Historical Demography" for the *Handbook*, found on pages 91–98. He recites the same data as found in his previous studies—that of 150,000 Indians in the state in 1845, a reduction to 100,000 in 1850 and another reduction to 50,000 by 1855. When the data is compiled from the journals that appear in this study, it becomes obvious that Cook's analysis is faulty.

the value of lands for agriculture, cattle raising, and other purposes, such as the extent of timber and coal. The journals and maps originating from the expeditions that General Smith put in motion, some of them over fifty pages long, have long been buried in the files of the National Archives and remained largely unused by scholars.[7] They are here reproduced. They offer vast amounts of new information on the environment, diseases (among both the Native and white populations), resources, geography, and early settlement in California.

Given the need, Smith had called upon the War Department to send out topographical engineers who could survey the region and create reliable, accurate maps. A few were sent, and they contributed markedly to the growing knowledge of California, but Smith also turned to regular army officers who could at least draw sketches of various regions. Whenever an expedition went out, the officer in command was expected to turn in a carefully written and complete report as well as a map. Three engineers arrived with the 2nd Infantry in spring 1849 who had been trained to author just such reports. They were Lieutenants George Horatio Derby, Robert Stockton Williamson, and William H. Warner. Another engineer by training, Lieutenant (later brevet major) Edward O. C. Ord, came with the 3rd Artillery. Other officers, who often tried to recruit assistance in making maps, joined them in surveying the vast expanses of California, including Lieutenants Nathaniel Lyon and Henry Walton Wessells and Captain Erasmus Darwin Keyes. Of the latter three, two of them promised to make maps that were never actually completed, or at least have disappeared from the army records of their surveys (the maps were often used during later expeditions and may have simply not been returned to the files). Nevertheless, Derby, Williamson, and Warner, who worked together; Lyon; and Ord all completed rather extensive maps of parts of California that are reproduced in the following pages.

[7] The massive collection of documents in NARG 393 contains thirty-one feet of "Letters Received" correspondence just for the period 1849–65 for the Department and Division of the Pacific. Seven of the journals reproduced in this book are found in this collection. A second, smaller collection of significant importance for the very early period of California military history is NARG 393, 10th Military Department, 1849–51. It has three feet of records in its "Letters Received" file. Both collections have other files, including massive volumes that hold "Letters Sent" files.

Lieutenant Derby was the most prolific, authoring four major journals of various expeditions in California and finishing four major maps. Best known today for his humorous sketches, Derby was born on April 3, 1823, in Dedham, Massachusetts. He graduated from West Point in 1846 and was assigned to the 2nd Infantry, arriving in California in early spring 1849. Derby joined Colonel Bennett Riley, then military governor of the region, on a survey of the new Mining District in late summer 1849. Traveling up most of the major river valleys south of the American River, Riley and Derby concluded that most of the Indians of that region were moving south, away from the miners, into the Kern and Tule river valleys. Derby estimated that their total population was only about 4,000 people, far fewer than general estimates reported in newspapers.[8] Derby's journal and map of the region also offered considerable information on the "diggings" that were under way, the various types of placer mining, and the tent towns that had sprung up throughout the mountains.[9]

Over the course of the next year, Derby struck out on his own, surveying the Sacramento River valley, producing an exquisite map of that region and, a few months later, turning to southern California, or the "Tulares Region," as it was called, providing another wonderful map and journal. On the upper Sacramento River, once again he found far fewer Indians than previous reports had suggested; the total Native population of the regions north of the American River and east of the Sacramento River was only about 4,000, he concluded. A terrible

---

[8]Cook puts the total Native population in 1850 at 100,000 Indians, most observers at the time suggesting that half that number lived north of a line from San Francisco eastward to the American River, and half south of that line. The first two Indian commissioners, designated by the Indian Office to negotiate with Natives upon arrival in California in 1851, put Native populations in the hundreds of thousands. They were reading the local newspapers, which generally reported such numbers. These numbers were considered somewhat reliable by Cook. See George Barbour and Oliver Wozencraft to Commissioner of Indian Affairs, March 5, 1851, NARG 75, LR, California Superintendency. Both commissioners quickly realized how inflated those early reports were and, after meeting Indians, gave actual counts that were nearly identical to Derby's assessment. See their later reports (ibid.).

[9]The original journal and map of the mining district are found in NARG 393, LR, DP. For a biographical sketch of Derby, see Wilson and Fiske, eds., *Appletons' Cyclopaedia of American Biography,* 2: 148. A full biography is found in George R. Stewart, *John Phoenix, Esq., the Veritable Squibob* (New York: Henry Holt, 1937), and a shorter, more interpretive piece is Canice G. Ciruzzi, "'Phoenix' Revisited: Another Look at George Horatio Derby," *San Diego History* 26 (Spring 1980): 1–12. Recently, Richard Paskowitz has published as e-books various pamphlets relative to Derby's life and his 1849 journal about the mining district. The publisher is listed as Lulu.com.

demographic decline was well under way; upon entering Native villages, Derby found that many Indians were dying of disease, most being afflicted with malaria.[10] Turning south, in 1850 Derby surveyed the San Joaquin River valley as well as the Tulares region, well to the south of it. Here, populations were suffering from diseases as well, particularly those villagers who lived close to the Tule Lakes. These lakes had massive swamps of many miles in length surrounding them, which were hatching grounds for mosquitoes, the vectors that carried malaria.[11]

Derby received orders to survey the lower Colorado River in late 1851. His task was to discover whether freight could be moved upriver from its mouth to support the newly created Fort Yuma, at the mouth of the Gila River. Emigrant trains followed the Gila River to where it struck the Colorado, crossing at that point into California. As with his earlier work, Derby succeeded not only in providing the necessary information but in convincing the high command to use small steamers for supplying the fort, one of which was on the river the next year.[12] Within a few years, there were four or five steamers plodding up the Colorado into Fort Yuma, and the region was opened to settlement.

All of Derby's journals offer extensive information on the geography, natural resources, and longitude and latitude of the regions visited. They also contain considerable information on the Native tribes, including population counts that for the first time were more than mere estimates or wild guesses published in newspapers. Derby's observations on the

---

[10] The original journal and map of the Upper Sacramento River Survey are found in NARG 393, LR, DP. Derby had a notorious wit, and his original journal contained passages that the army found unacceptable. Consequently, the original journal of his Sacramento expedition has sections that were edited out when it was published in the serial set. The edited version is reproduced in Senate Executive Document 47, 31st Congress, 1st session, part 2, 1850, serial 558. Given the immense interest in the California gold strike, document no. 47, which contained many more military reports from California, was published in Tyson, *Geology and Industrial Resources of California*. Derby's report on the Sacramento was also reproduced as Francis P. Farquhar, "The Topographical Reports of Lt. George H. Derby," *Quarterly of the California Historical Society* 11 (Spring, Summer, Fall 1932).

[11] The journal and map of Derby's expedition into Southern California are found in NARG 393, LR, DP. They were reproduced in Senate Executive Document 110, 32nd Congress, 1st session, serial 621.

[12] The original copy of Derby's Colorado River journal is not in the Department of Pacific Papers in the National Archives. It was published in Senate Executive Document 81, 32nd Congress, 1st session, 1852, serial 620. It was also reproduced as *Derby's Report on the Opening of the Colorado, 1850–1851*, ed. Odie B. Faulk (Albuquerque: University of New Mexico Press, 1969).

gold country and the Tulares region, sources virtually unknown to historians, are exceedingly valuable. While Derby's humor has been celebrated and frequently cited and reprinted, his journals are largely overlooked, found today in their original form in the War Department files. When his extensive maps are added, they offer a unique perspective on California, one missing from the history books.

One of Derby's early biographers, Francis P. Farquhar, sensed that Derby lived a "double Life." Readers know him as a major humorist whose many articles in California newspapers during the 1850s entertained the state. His father, a poet of sorts, obviously had encouraged such expressionistic writing. But while at West Point, although known for dry humor, Derby became a serious scholar, studying engineering and graduating in 1846. While Farquhar concluded that Derby's most important scientific contribution was his journal of discovery along the Colorado River, this assessment is confined to Farquhar's generation of historians. Today, Derby's work is monumentally important to Indian historians, who struggle to come to some understanding of what truly happened to the Indians of California. Farquhar was not at all interested in the subject; indeed, Cook's work on Indian demographics was still thirty to forty years in the future. Perhaps Derby himself never understood his achievements in this regard, as he spent the last nine years of his army career working on river diversion projects and lighthouses.[13]

Derby's time in California was not always notable for demonstrating smooth relations with other, more senior officers. He had a flippant, enigmatic personality that led to occasional confrontations. His journal for the Sacramento expedition offers a case in point. When reaching the mouth of the American River, he described the new town of "Sutterville," a place on the Sacramento River "where a large city has been projected, but which at present is occupied by a few hastily constructed shanty's." The inhabitants, he claimed, "are called Suttervillians, whether appropriately or not I am unable to say." He offered yet a second critique of another city along the Sacramento River: Derby thought its potential for becoming a "large commercial place"

[13]Farquhar, "The Topographical Reports of Lt. George H. Derby," 99–100.

was limited due to the "trifling disadvantage of [it] being under water for four months out of the year." Such comments riled senior officers, and the journal, once at headquarters, had several sections marked "omit." In other words, while the original journal is here reproduced in its entirety, in the two cases where segments of the journals were reproduced in the serial set, sections have been omitted.

Yet a second brush with senior officers occurred in 1850 after a soldier stole his horse. Derby called for the arrest of the man, only to discover that a local court had captured him and prepared to execute him. Derby abhorred the court's decision and helped the man escape. This brought retribution from Lieutenant Colonel Joseph Hooker, who would obtain his own fame and notoriety during the Civil War. Hooker charged Derby with aiding a deserter from the army. The court-martial of the young officer took several months, and Hooker eventually moderated in his quest for punishment. (Perhaps he recognized the value of Derby's talents.) Regardless, Derby received a reprimand and resumed his duties as head of the Topographical Corps in California.[14]

Derby's later career, after his last major survey in 1852, involved purchasing a newspaper in San Diego and the publication of numerous satirical articles. Penned under the pseudonyms "John P. Squibob," a nickname that he acquired at West Point, and, later, "John Phoenix," these satires became immensely popular in California. They involved such exotic topics as "Lectures on Astronomy" and the infamous attempt to create "A New System of Grammar," based on numbers that described words (e.g., 100 was assigned to the word "perfection").[15] The growing popularity of Derby's essays resulted in a book entitled *Phoenixiana: or, Sketches and Burlesques,* published under the name John Phoenix.[16] Derby made little money from the venture and took sick a few years later, likely from a brain tumor. Briefly returning to duty in Florida, he died on May 15, 1861, just as the American Civil War began.

[14]The only major biography of Derby is Stewart's *John Phoenix.* His court-martial is discussed on pages 79–92.

[15]Many of his articles, which became extremely popular in California, were published in the *San Diego Herald* in the early to mid-1850s.

[16]The publication is titled John Phoenix, *Phoenixiana; or, Sketches and Burlesques* (New York: D. Appleton, 1856). It too was popular, going through thirteen printings.

Edward O. C. Ord was one of the most experienced officers who worked in California. Born in Maryland in 1818, he graduated from West Point in 1839 with an engineering degree.[17] Assigned to the 3rd Artillery, he reached California in 1848 after serving in both the Seminole and Mexican Wars. Like fellow classmates William Tecumseh Sherman and Henry Halleck, who also were in California, Ord had a long, distinguished career. Remarkably, little is known about his survey work. Most biographies note his service in both the Rogue River War of 1856 and the Spokane War of 1858. During the Civil War, Ord joined Ulysses S. Grant in Tennessee and was on hand at the fall of Vicksburg. He also helped force the surrender of Robert E. Lee in April 1865, rising to the rank of lieutenant general and commanding a corps. Ord retired from active service in 1880, turning to railroad construction. He died of yellow fever in Havana in 1888.

In California, Ord is known as one of the first men to explore and map the Los Angeles basin. A copy of his 1849 map of the basin, and early Los Angeles, is found in the city clerk's office. Little known or cited by historians are Ord's two early journals of his 1849 expedition, which have survived in the National Archives and provide basic information about the town itself as well as the ranching community along the coast. Originating from Mexican cessions in the 1830s and 1840s, these ranches had vast numbers of cattle and horses but produced very few crops. While the ranches had once been sources for cattle hides, this industry had been abandoned by the time Ord arrived, the cattle having been moved into the mines and sold. One of Ord's charges was also to determine what passes existed in the mountains to the east of Los Angeles that might be used by Indian raiders. General Smith eventually built military posts to secure the passes explored by Ord. His journals recounting these efforts are published here.[18]

---

[17]A biographical sketch of Ord is in Wilson and Fiske, eds., *Appletons' Cyclopaedia,* 4: 417.

[18]The two journals and his map of the Los Angeles basin are in NARG 393, LR, DP. They are reproduced in Senate Executive Document 47, 31st Congress, 1st session, part 2. Ord's 1856 journal of yet another visit to the Los Angeles region is in the Henry E. Huntington Library and has been published as *The City of the Angels and the City of the Saints,* ed. Neal Harlow (San Marino, Calif.: Huntington Library, 1978). The map used in this later publication is from his 1849 expedition. We are indebted to Peter Blodgett at the Huntington for helping us sort out the Ord material.

Much can be learned about the early California coast from the Ord journals. He offered wonderful descriptions of the various missions, from Monterey south to San Diego—his journals even specify the number of adobe houses in the town. Some of the missions were still intact, complete with their libraries. Ord, likely the first serious preservationist in California history, suggested the need to quickly find someone to "put in charge" who would see to the missions' "preservation." Ord spent time with many of the leading Mexican citizens of the Los Angeles basin, including Don Bernardo Yorba and Don Pío Pico. Both had extensive ranches with vineyards that had, given the war, fallen somewhat into decline. Ord surveyed the three major rivers of the basin, the Santa Ana, San Gabriel, and Los Angeles, and he examined the passes over the mountains to the east, in particular Cajon Pass. Finally, he entered Los Angeles itself, described its multiethnic population—with some degree of humor—and noted its one hundred "adobe houses." There is likely no better description of the region available, for the late 1840s, than Ord's accounts.

Young Lieutenant Robert Stockton Williamson graduated from West Point in 1848, two years after Derby. The two young officers obviously knew each other at school. Williamson, also assigned to the topographical engineers, reached California in 1849.[19] He showed considerable grit that fall when, after the murder of his commanding officer, First Lieutenant William Warner, by Indians, Williamson continued the expedition into northeastern California, completing a journal and map of the region that Warner had begun. General Smith was particularly impressed with the young man, and Williamson continued on in California for two years, surveying various locales for the army and producing maps.[20] But his journal and map of northeastern California, including careful surveys of both the Pit River and Goose Lake, remain his most significant contributions and are reproduced here from the originals.[21]

---

[19]A biographical sketch of Williamson is found in Wilson and Fiske, eds., *Appletons' Cyclopaedia*, 6: 537–38.

[20]See Smith to Lieutenant Colonel William G. Freeman, February 2, 1850, NARG 393, LR, DP.

[21]See Williamson report, February 14, 1850, NARG 393, LR, DP. The report is reproduced in Senate Executive Document 47, 31st Congress, 1st session, part 2.

What is striking about Williamson's journal is the fact that the party, after many weeks in the northwestern mountains of California, encountered virtually no Indians, not even the war party that ambushed Captain Warner, which had apparently moved back north. There was evidence of Native occupation, in the form of abandoned traps used to take bears and other animals, but no signs of villages or even fishing weirs. The Williamson party did spend time with Peter Lassen, the early settler who had opened another trail from the Humboldt River into northern California, but the party concluded that the mountain passes, which Lassen had suggested were usable, were not capable of handling railroad traffic. Indeed, Williamson decided that the northwest mountain region was literally too rugged to ever be populated, by Indians or whites.

Williamson went on to work on railroad surveys in 1853, was promoted to first lieutenant three years later, and served as chief of the topographical engineers for the Army of the Potomac during the Civil War, rising to the rank of lieutenant colonel. While he completed many important maps of Virginia for the army during the war, even going behind Confederate lines on occasion to acquire information, Williamson suffered from tuberculosis after the war and eventually retired to San Francisco. Mount Williamson in California and the Williamson River in Oregon are named for him. He died in California on November 10, 1882.

After Williamson's return from Pit River and Goose Lake, the army sought to "chastise" the Indians responsible for several murders, including those of Lieutenant Warner as well as two ranchers near Clear Lake. Senior officers selected Captain Nathaniel Lyon for the task. Lyon graduated from West Point in 1841, then went immediately to Florida where he served with the 2nd Infantry against the Seminole Indians. He fought valiantly at Veracruz and Churubusco, where he was wounded, and gained the respect of many senior officers.[22] Given his experience and demonstrated bravery, Lyon was selected to lead the first two major expeditions against Indians in California, destroying several Pomo Indian villages north of Sonoma but failing to punish

[22]A biographical sketch of Lyon is found in Wilson and Fiske, eds., *Appletons' Cyclopaedia*, 4: 67–69. A biography that focuses on his Civil War career is Christopher Phillips, *Damned Yankee: The Life of Nathaniel Lyon* (Columbia: University of Missouri Press, 1990).

the Indians responsible for the death of Lieutenant Warner, who was killed in the northeast, near Goose Lake. He left an initial report of his attack on the Pomo Indians, followed by a much more detailed journal as well as a final journal of his expedition into the Pit River valley of northeastern California. These expeditions are exceedingly important accounts of the early history of California, but have never before been published.[23]

Lyon's journals reveal a number of important facts about the geography of Pomo Indian country and, for the first time, describe the terrible retribution that they suffered at his hands. The journal demonstrates once again that reports of thousands of Indians living along Clear Lake, as well as near the Russian River, were complete exaggerations. The Pomos from the northern half of Clear Lake could muster only about fifty men to oppose the army. Lyon used dragoons to force them onto an island, then launched boats on the lake to reach the island. The Pomos attempted to resist and "howled the long night, in fierce denunciation & bold defiance," but they were "slaughtered" when the troops attacked. Worse, Lyon brought even greater devastation to yet a second village on the Russian River, one of only two that existed there. Most historians have heard of the defeat of the Pomos at "Bloody Island," but the attack on the second village is unknown to California history. It was more destructive than the first assault on the island and likely constituted the most violent campaign ever launched by the army against California Indians.

Captain Lyon learned much about the regions north of Sonoma and the lands that bordered the upper Pit River. A few venturesome ranchers had moved into the Napa valley and to the north of it, two or three receiving Mexican land grants just before the war broke out. They needed labor and, at times, apparently kidnapped Indians for that purpose. It is uncertain whether Lyon knew of these prior abuses; he had orders to "chastise" the Indians, and he followed them. He also learned, however, that "Spaniards," or Mexicans, might have been involved in urging the Indians to kill American ranchers. While he was unable to find conclusive evidence to that effect, it is obvious from

[23]Both journals, along with the map reproduced with them, are found in NARG 393, LR, DP.

Lyon's journal that much was going on north of Sonoma that led to violence, and the region would continue to be unstable for the next dozen years, as conflict spread into the Russian River and then eastward into the Eel River.[24] While Lyon attempted to find the Indians responsible for Warner's death near Goose Lake—they were likely Klamath or Modoc warriors—these people lived well north of the California border. Lyon's journal of his campaign into northeastern California, much like Williamson's, only confirms that this region of California was virtually deserted.

Lyon had a reputation for being a firebrand, which may explain why he was selected to lead the expeditions against Indians that the army viewed as "hostile." After returning east, he served in Kansas during the difficult mid-1850s, quickly developing sympathy for the Free-State Party. He commanded federal troops in St. Louis at the beginning of the Civil War, refusing to allow southern sympathizers to gain control of the federal arsenal there. President Abraham Lincoln personally selected Lyon to raise a volunteer army to defend Missouri, sending in several Illinois regiments to support his efforts. Recruiting mostly German immigrants, Lyon moved on Jefferson City, which he captured, and then faced Confederate troops under Generals Sterling Price and Ben McCulloch at Wilson's Creek in early August 1861. While outnumbered two to one, Lyon attacked, was shot from his horse, found another mount despite suffering from several wounds, and ultimately was killed in the final assault. He is often credited with saving Missouri for the Union cause in the war.[25]

Lyon became one of the first martyrs to the Union cause, being virtually deified within months of his death. He had turned to writing in the late 1850s, addressing the issues of slavery and abolitionism. His writings were quickly gathered by an ambitious editor and published as *The Last Political Writings of Nathaniel Lyon, USA; with a Sketch of His*

---

[24]The conflict has been described as "genocide," although the number of Indians in the region did not range into the thousands. Actual reports from the battlefield suggest that Lyon's forces killed two or three hundred Indians. For the discussion of genocide, see Benjamin Madley, "California's Yuki Indians: Defining Genocide in Native American History," *Western Historical Quarterly* 34 (Autumn 2008): 303–34.

[25]See Phillips, *Damned Yankee.*

Life and Military Service.[26] Although Union soldiers failed to retrieve his body after the battle at Wilson's Creek, it was later interred in his hometown, Eastford, Connecticut. Some fifteen thousand people attended the funeral, and Lyon's name was soon attached to streets, and even counties, many newly created in Minnesota, Iowa, Kansas, and faraway Nevada. Yet little is often said of the terrible retribution that he put upon the Pomo Indians, some of whom were not guilty of killing anyone, as his journal from the California campaigns suggests.

Yet another experienced officer who became part of General Smith's survey of California was Captain Erasmus Darwin Keyes, an 1832 West Point graduate. Assigned to the 3rd Artillery, Keyes was sent to Charleston Harbor at the height of the Nullification Crisis that year. He then joined General Winfield Scott during the forced removal of the Cherokee Indians in 1837 and 1838. Keyes reached California in 1848 and was assigned escort duty to the Indian commissioners who negotiated with the California tribes of 1851. His journal, kept during that summer, offers considerable insight into the various bands that inhabited the river valleys coming out of the Sierra Nevada, from the San Joaquin River southward to the Tulares region.[27]

Keyes is particularly careful with population estimates. At one point he gives the total Native population for the Four Creeks region as "1,240 individuals," suggesting an actual count. When all the Indians that he saw and dealt with are added up, the total population of those along the rivers from the San Joaquin south amounted to a little over 4,000, or the same number that Derby arrived at. Keyes also offers advice on locations for army posts, the quality of the land, some of it "poor," while other regions were quite good. While a short memoir, Keyes's journal, once again, reaffirms the careful demographic work done by the army in determining the location and numbers of Natives in California.

Keyes, like Ord, fought against Indians in Oregon and Washington in the years 1855–58. When the Civil War broke out in April 1861, he became brigadier general of volunteers. The next year he commanded

---

[26]See Lyon, *The Last Political Writings of Nathaniel Lyon,* USA*; with a Sketch of His Life and Military Service* (New York: Rudd & Carleton, 1861).

[27]A biographical sketch of Keyes is found in Wilson and Fiske, eds., *Appletons' Cyclopaedia,* 3: 530.

a corps and fought in most of the Virginia campaigns, including Lee's Mill, Yorktown, Bottom's Ridge, Savage's Station, and Seven Pines. Retired from active service in 1863, Keyes returned to California, speculated in gold mining, and became wealthy. He then turned to writing, publishing three books, *The Rear Guard at Malvern Hill* (1884), *Fifty Observations of Men and Events* (1884), and *Fighting Indians in Washington Territory* (ca. 1884).[28] He died in 1888 at the age of eighty-five while on a vacation to Europe.

One of the least known regions of California was the land north of Sonoma, comprising the Russian, Eel, Klamath, Trinity, Scott, and Salmon river valleys. Mountainous and almost inaccessible by wagon roads, these regions were believed to have thousands of Indians, most of whom were thought to be antagonistic toward whites. Captain Henry Walton Wessells led the first major military expedition into these lands in the summer and fall of 1851. Wessells graduated from West Point in 1833, a year after Keyes, and the two men had remarkably similar careers in the years that followed. Wessells fought in the Seminole Wars between 1837 and 1840 and then in Mexico, where he was cited for gallantry at Churubusco and brevetted to major. He reached the Pacific Slope in 1849, joining Indian Commissioner Redick McKee on escort duty into the wilds of northwestern California two years later. His journal of this expedition is one of the most important ever recorded for the early history of that portion of the state and its people.[29]

Wessells and McKee spent several months traveling through the northwestern regions of California, lands that necessitated the use of mules for transportation. They learned quickly that the Native populations of those regions were quite small, and that almost all Indians lived along the river valleys. Wessells counted them as they traveled northward, some 700 at Clear Lake, only a few hundred more on the Russian River, and some 900 on the central Eel River. These numbers were small, and as Wessells notes, most Natives were "loathsome from

[28]See Keyes, "The Rear Guard at Malvern Hill," in R. U. Johnson and C. C. Buel, eds., *Battles and Leaders of the Civil War* (New York: Century War Series, 1884–87), *Fifty Years' Observations of Men and Events, Civil and Military* (New York: University of Michigan Library, 1884), and *Fighting Indians in Washington Territory* (reprint, Fairfax, Washington: Glen Adams, 1988).

[29]A biographical sketch of Wessells is in Wilson and Fiske, *Appletons' Cyclopaedia*, 6: 438–39.

disease." The party reached Humboldt Bay, found a thousand or so Indians where the Klamath and Trinity Rivers unite, then proceeded to the Scotts River, and finally headed over the mountains into the Sacramento valley. Wessells was a keen observer who noted the clothing that Indians were wearing and made comments regarding their economies and even their burial customs. He also discovered and noted the first appearance of American miners in these valleys, men who would later force many of the Native villagers out, either onto newly created Indian reservations or into the mountains, to starve.

Wessells's career continued to involve frontier service after he left California. He marched with General William S. Harney on the Sioux Expedition of 1855. As the Civil War broke out, he became colonel of the 8th Kansas Regiment, and soon thereafter brigadier general. Wounded at Seven Pines while fighting alongside General Keyes, he was sent south to help defend North Carolina. There, he was forced to surrender, was later exchanged, and served out the remainder of the war in the North. He retired from the army in 1871 and returned to his hometown of Litchfield, Connecticut. He died quietly in 1889.

All six of these army journalists, included in this collection, shared a dedication to duty, were brave in battle, and possessed obvious intellectual capabilities. Five of them—Lyon, Wessells, Ord, Keyes, and Derby—had seen combat against Indians and Mexicans and were wounded during action in the Mexican War. The five were cited for bravery during these engagements, Wessells picking up the regimental flag that had fallen and using it to urge his men forward. Litchfield, Connecticut, his hometown, later presented him with a sword for gallantry. Lyon, Derby, Keyes, and Williamson all continued to write after their early journal efforts in California. Williamson's works were likely the least known given that his efforts were reports on the feasibility of railroad routes in the West. Not all their achievements were published; Ord and Williamson mapped and plotted the street grids for the town of Sacramento in 1849, giving it the uniform organization that it enjoys today.

Several of these men came from upper-class families. Williamson got into West Point on the recommendation of family friend Commodore

Robert F. Stockton, whom Williamson was named after, and whose naval units helped bring California into the Union. Stockton later became a United States senator.[30] Derby came from a well-off family, his father abandoning business at an early age to take up the writing of poetry. Edward O. C. Ord had the most interesting, and mysterious, background. His father, James, was reputed to be an illegitimate son of King George IV of England. Ord was considered a mathematical genius. While Keyes's father was more common, he was a noted surgeon in his time. And even Wessells and Lyon had early schooling, which, along with family connections, was absolutely necessary in order to gain admission to West Point.

Thus, all six of these journalists were well-educated, somewhat elitist individuals whose descriptions of life and people offer wonderful insights into the lands that soon thereafter became a prosperous state, but who came from a class standing that, at the time, often proved discriminatory and unbending. Accordingly, as journalists, these men often offered unsympathetic and unflattering descriptions of people, observations that they viewed as accurate—but descriptions that the army often omitted from reports sent east. But they were hardly alone in these biases, as the works of the great liberal writer Mark Twain later demonstrated. In *Roughing It,* his wonderful portrayal of early California, Twain also found few people in the state worthy of his attention and sympathy, especially the Native inhabitants.[31]

Like Twain, the journalists often had disparaging words when it came to Native Americans and the elements of Native cultures that they observed. Indeed, they often adopted the common usage of the time, identifying California natives as "diggers," a reference to their efforts to search for roots and herbs. These men also came upon the California tribes at a point where miners were occasionally dislodging them from their salmon fishing sites or destroying the salmon runs, and when massive epidemics raged through California, carrying off

[30]For a biography of Stockton, see Wilson and Fiske, eds., *Appletons' Cyclopaedia of American Biography,* 5: 694–95.

[31]Twain went to Nevada in 1861, moving to San Francisco several years later. His humorous rendition of events was published eleven years later. See *Roughing It* (Hartford, Conn.: American Publishing, 1872).

thousands of Indians. Little wonder that the survivors often looked emaciated, poorly clothed—or often without clothes—and prone to begging for food. To be sure, these six journalists often were just as harsh in their descriptions of *Californios*, or Mexicans, and even many emigrant families who arrived in California, broke and worn out from the long trip.

The attitudes and biases that are revealed in these early journals, then, come with the age. It was an exciting age, one of discovery and the opening of new land, when viewed from the white perspective, but one of distress and the loss of land, when viewed from an Indian or even, perhaps, from a Mexican perspective. The six journalists whose efforts appear here understood that bias to some degree, and in that sense their contributions should be viewed like those of such earlier explorers as Lewis and Clark, Zebulon Pike, Stephen Long, or even John C. Frémont. As primary sources, these journals reveal a land at the point of contact, or what is often called a "frontier" stage, where Natives and non-Natives observed one another and often clashed with each other. The journalists saw their roles as that of promoting the peace when possible and gaining the knowledge of those lands that General Smith thought essential. In that regard, their missions were certainly a success.

CHAPTER I

Derby's and Ord's First Journals: The Mining District and the Basin of Los Angeles 1849

George Horatio Derby's Journal of the Mining District[1] July 5–August 11, 1849

As the professional army entered California in early 1849, it quickly realized that the concern of earlier volunteer troops in the region—the possibility of Mexico retaking the region—had disappeared. Two concerns seemed most apparent to General Persifor F. Smith and the young officers who served with him: to protect the mining district and the gold that was coming from it and to stop the raiding parties of Indians who had become so problematic for the Californios and

[1]The original copy of Derby's journal is found in NARG 393, LR, DP. While Governor Bennett C. Riley, who led the expedition, forwarded several reports to Washington briefly describing the results of the trip, and his report dated September 20, 1849, included a copy of Derby's map of the mining district, Derby's original journal was never published in the serial set. For Riley's many brief reports, see *House Executive Document 17*, 31st Congress, 1st session, 1850, serial 573. Riley's reports for September 20 (pages 941–45 of the serial set), and his report for October 15, 1849, which was not included in the serial set publications, are published here.

Bodaga
Sonoma
King's Point
PABLO BAY
SULUN BAY
Benicia Camp
Martinez
Aug. 2
New York
BAY of SAN FRANCISCO
SAN FRANCISCO
Point San Pedro
Mission
Livermore
Santa Clara
San Jose

"Sketch of General Riley's Route through the Mining Districts, July and Aug 1849."
John McHenry Hollingsworth, in the office of the 10th Military Department, made this copy of a map originally created by Derby in his journal of the mining district, July–August 1849. *Courtesy of National Archives and Records Administration.*

their ranches, the most important being in a district that extended northward from San Diego into the Los Angeles Basin. Two officers were sent to examine these lands, determine the strength of Indian populations, and construct maps. They were Brevet Lieutenant George Horatio Derby and Captain Edward O. C. Ord. General Bennett Riley, who led the party that Derby was attached to, added two brief reports to Derby's final journal, giving his assessment of Indian affairs in California.

Orders to Lieutenant Derby

Camp near Benicia
Lieut Geo H. Derby T.E. June 17th 1849

Sir

You will please repair without unnecessary delay and report in person to General Riley for this purpose of accompanying him on a tour through the San Joaquin & Sacramento vallies.[2]

If you can procure a "Schmalcalder" compass you are authorized to buy one or a good common pocket compass or both. The "Schmalcalder" I would recommend the use of as being much the most accurate.[3]

It is desirable that you should gather all the information possible of any kind that you may think will be interesting and useful to the Topographical Department.

When you have completed the tour (unless otherwise directed) you will report in person at Head Quarters of the Topographical Department in California wherever they may be for the purpose of making a map of the country you will have passed over. You can doubtless procure note books in San Francisco suitable for your purpose.

---

[2]General Riley was the governor of California from 1849 to 1850, after which statehood was granted by Congress.

[3]The Schmalcalder compass was invented by Charles Augustus Schmalcalder, from Germany, and patented in London on March 5, 1812. The compass patent was a refinement of an earlier compass in that Schmalcalder added a folding prismatic sighting mechanism with sighting vane. The compass is similar to the common military compass still used today by military organizations around the world, including in the United States Army.

I would recommend that you make a field plat of your notes as you advance. This, if done daily, will be an improved duplicate of your original notes.

I am very respectfully
Your obedt Servt
(Signed) W. H. Warner
Brevet Captain Topographical Engineers

To Maj E. R. S. Canby[4]
Adj. Genl 10th Mil Dept. Cal

Sir

In compliance with the above order received by me from Capt Warner on the 19th of June last, I had the honor of reporting to you in person at San Francisco, which place I left with the commanding General [Riley] on the 23rd in the *Mary Jane*, arriving at the Head Quarters of this Department [Department of the Pacific at Monterey] on the 25th.

I had endeavored to procure such instruments as I considered necessary on a tour like that contemplated in San Francisco, but finding it impossible to do so, I was compelled to leave entirely unfurnished, with the exception of a Schmalcalder compass for which I was indebted to the kindness of Capt Halleck of the Corps of Engineers.[5]

We left Monterey on the 5th of July, the party consisting of the Commanding General [Riley] & Staff. Sergeant Carson of the Artillery

[4]Major Edward R. S. Canby graduated from West Point in 1839. He served in the Seminole and Mexican Wars before being assigned to the California frontier. He later rose to the rank of general during the Civil War, mounted a thoughtful defense of New Mexico that thwarted Confederate designs, and was active in the capture of the lower Mississippi River. He tried to negotiate a peace with the Modoc Indians in 1873 and was assassinated by them during a council. Canby was the only general officer to die during the Indian wars.

[5]Captain Henry W. Halleck, later a general in the Civil War, graduated from West Point in 1839, the same year as Canby. While in California, Halleck collected massive numbers of Spanish documents, mainly from the missions, which were later deposited at the Bancroft Library, Berkeley, California. Halleck played a major role in the Civil War, orchestrating the western theater where Ulysses S. Grant won major victories while under his command. Halleck died in 1869, while still in the army.

as guide, Sergeants Bingham and Ferry of the Dragoons, & Privates Newman and McNab as muleteers, and five pack mules carrying our provisions etc.[6] Marching nearly N.N.E. & crossing the Sandy Hills which form the line of coast in the vicinity of Monterey we came to a shallow stream known as the Salinas or Monterey River, which rising in the spurs of the Coast Range, runs in a Northwesterly direction and empties into Monterey Bay some ten or fifteen miles from the town—fording this, the water being but about a foot in depth, we came out on an extensive prairie, covered with fine grass, which extends to the range of hills in the vicinity of the Mission of San Juan [Bautista].[7] Crossing this prairie & leaving the usual encamping ground a mile in the rear we took up our quarters for the night at the entrance of the pass through the hills, where we found good water, & an abundance of fine trefoil[8] usually termed here "Antelope Clover" with which the ground was completely matted, affording excellent grazing for our animals. The ground at the base of the hills is thickly timbered, principally with the holly oak & sycamore. We were surrounded at night by a dense fog arising from the valley which condensing towards morning fell like a shower of rain wetting thoroughly everything that had been accidentally left exposed. We left our uncomfortable encampment at 20 minutes past 7 on the following morning and marching nearly East crossed the small stream which had supplied us with water, and struck the defile between the hills which rise on some parts of the range to

[6]Sergeant James H. Carson arrived in California with the army in January 1847. He left three memoirs detailing an early history of the gold rush, including "Early Recollections of the Mines," "Tulare Plains," and "Life in California." They were published three times, most recently as *Bright Gem of the Western Seas: California, 1846–1852*, ed. Peter Browning (Lafayette, Calif.: Great West Books, 1991). One of the earliest publications is *Recollections of the California Mines: An Account of the Early Discoveries of Gold, with Anecdotes and Sketches of California and Miners' Life and a Description of the Great Tulare Valley* (1853; reprint, Oakland, Calif.: Diobooks, 1950). Carson died in San Francisco in 1853. Where individuals such as Bingham, Ferry, Newman, and McNab cannot be identified, there is no footnote.

[7]Located just northeast of Monterey, Mission San Juan Bautista was built in June 1797, the church being finished by Christmas of that year. By 1820, about one thousand Indians lived near the mission, and it possessed a cattle herd of over six thousand head. While about half the missions in California were secularized by the Mexican government in 1834, San Juan faced that change the next year. See Irving Berdine Richman, *California Under Spain and Mexico, 1835–1847* (New York: Cooper Square Publishers, 1965), 254–55.

[8]Trefoil is any of the various plants of the pea family having three leaflets to each leaf like the clover, black medic, and tick trefoil.

the heights of six hundred feet on either side.[9] The first part of the defile we found pleasant. The road, easy and surrounded on every side by willows, holly and post oak, with wild roses and other flowers in profusion, winding around the base of larger hills we came at length to the last of the range, which the road crossed. It was excessively steep and I should suppose scarcely practicable for wagons. We scrambled up however with our pack mules meeting with no inconvenience but the loss of a horse belonging to one of the party, which gave out from exhaustion. From the summit we caught through the dense and chilling fog glimpses of the beautiful valley of San Juan which extends through from the main chain of the coast range to San Francisco. At the foot of the declivity we observed a large and apparently well cultivated rancho, with many cattle, sheep, and goats in its vicinity. I did not learn the name of the owner.[10] These spurs of the coast range which we have just crossed run nearly N.N.W. They are of course of volcanic formation, exposing at different points on our route large outcrops of gneiss, trap & basalt and near the rancho I observed one hill the entire side of which appeared to be fine white limestone.[11]

We observed [at] the mission of San Juan, a small collection of tiled adobes about a mile N.E. from the foot of the pass. There are now six families of Americans residing there in addition to the native population which amounts to in all some one hundred and fifty or two hundred. We halted some little time for the mule to come up when striking a little N. of E. moved across a large prairie thickly covered with fine & excellent grass upon which large herds of cattle were feeding; in the direction of Mr. Pacheco's rancho which is situated at the base of the main chain of the "coast range" at the entrance of what is termed Pacheco's Pass, a gorge through which the road passes to the San Joaquin Valley. Pacheco's Pass is marked in the distance by a remarkable peak which being covered with trees set out dark in

[9]A defile is a narrow passageway through which troops pass in single file.

[10]Most of the land attached to Mission San Juan was given out in Mexican land grants, starting in 1839. See Burgess M. Shumway, *California Ranchos: Patented Private Land Grants Listed by County* (San Bernardino, Calif.: Borgo Press, 1988).

[11]Gneiss is metamorphic rock composed of quartz, feldspar, and mica. Trap rock is basalt or other fine-grained, dark igneous rock having a more or less columnar structure. Basalt is a hard, dark volcanic rock.

bold relief against the horizon in contrast to the low hills forming the commencement of the range which are entirely barren or covered with fine dried grass.[12] In leaving the pass at San Juan for Pacheco's this peak should be kept a little to the right, as though making for its base. We were received with much hospitality by Mr. Pacheco, an elderly man with a family consisting of a wife, a son and two daughters.[13] He informed me that he had thirteen thousand head of cattle feeding upon his rancho and five or six hundred horses. The distance from San Juan is about twelve miles. There are two or three long adobe buildings shaded by large & beautiful oaks & several huts for Indians etc. at the place. Leaving Mr. Pacheco's we proceeded in a N.E. direction to strike the entrance of the pass, crossing in the 1st mile or two the dry bed of a branch of the Pajaro River, several times, which however runs generally parallel with the trail & formed the bed of the valley through which we moved—finding a small quantity of water, in the bed, at a distance of 12 miles from Pacheco's accompanied with good grass in the vicinity.[14] We encamped for the night, having marched about twenty seven miles generally in a N.E. direction.

Our encampment was at the North base of a somewhat remarkable hill, about 600 feet high, the top cropping out with trap and basalt in fantastic shapes resembling somewhat the ruins of buildings at a little distance. Monterey bears from this hill about S.S.W. We were disturbed during the night by a party of furloughed and discharged soldiers on their way to the mines, who were prowling about the camp for some time with their animals before they could find the water. The next morning being Saturday July 7th we marched at seven o'clock crossing the bed of the stream again within ¼ of a mile of our encampment, and half a mile farther to a tremendously steep hill over which ran the footpath and up which our mules toiled with great difficulty.

[12]Pacheco Pass, along California Highway 142, separates the coastal region from the Central Valley. It was named for Francisco Perez Pacheco and was used by Yokut Indians to move back and forth between the coast and the interior for trade. See Erwin G. Gudde and William Bright, *California Place Names: The Origins and Etymology of Current Geographical Names*, 4th ed. (Berkeley: University of California Press, 2004), 233.

[13]Pacheco (1793–1860) received two grants for Rancho Ausaymas in 1833 and 1836 from Governor Nicolás Gutiérrez.

[14]Pájaro River, in Spanish, translates "bird." The river emptied into Monterey Bay.

After nine miles of ascent & descent in some cases almost precipitous we at length emerged upon a broad prairie some five miles in width by four in length enclosed by low hills, & thinly covered with buffalo grass, an unwholesome looking reed which I am told no animal but a buffalo can possibly digest. The coast range which we had now crossed extends from Oregon to Lower California interrupted occasionally as at [San] Pablo & Suzun [Suisun] Bay but sufficiently well defined on its whole line of direction which is nearly N.W. and S.E.[15] The spurs make off at different angles & at various points toward the sea. The hills vary in heights from to 2 [200] to 1200 feet and are sparsely covered as far as the eye could reach with the beautiful holly oak which dotting them in every direction with spots of brilliant green affords a pleasing relief to their otherwise brown and sterile appearance. The range is of course the results of volcanic action—I noticed two or three remarkable peaks with singularly shaped masses of trap & basalt crowning their summits, one of which looks precisely like an ancient church & might readily be taken for one, were it possible to erect it upon such a site or get at it after it was erected. No Gold or other minerals have as yet been found in the valleys of this range although I have no doubt that in some parts rich discoveries will be made. Our bearing in crossing the prairies was about N.N.E. I have begun to suspect the accuracy of the compass with which I had been furnished. The needle being rusty did not traverse well and I presumed had lost some of its power, an opinion in which I was the next day confirmed, by its making the sun rise in the North East with other like vagaries of an equally startling nature. Not caring to emulate the example of Joshua & dreading the effect of farther use of such an instrument, I put it carefully in my valise & judged the direction travelled by the position of the sun.[16] Having crossed the prairie we came to a lagoon about twenty feet wide of fine clean water which communicates during the rainy season with the San Joaquin River by a little stream at

[15]San Pablo and Suisun Bays form part of the large, then marshy, backwater northeast of San Francisco. Suisun Bay took its name from the Suisune, or "Suisun," Indians. Derby used the bays as reference points, noting how they were the only difficult geography in the Central Valley. For the Suisun, see *California*, vol. 8 of *Handbook of North American Indians*, 83, 350.

[16]Derby's reference is likely to Joshua 10:13: "So the Sun stood still, and the moon stopped, till the Nation avenged itself on its enemies, as it is written in the Book of Jashar."

present dry. A fine belt of holly oak encircles this lagoon & good grass covered this prairie in the immediate vicinity; for a party encamping at Pacheco's on their route to the San Joaquin this is doubtless their best point to make on the next days march as the ferries on that river are an easy days march from the Lagoon.

Having watered our animals, we crossed the bed of the lagoon some 3 [300] or 400 yards above and marching for twenty one miles in a direction nearly parallel to and about one mile from the easterly spurs of the Coast Range we passed between these singularly egg shaped hills & a most remarkable formation of trap rock in the shape of a regularly planned fortification, covering about an acre of ground & some ten or fifteen feet in height. It is almost entirely covered with a greenish yellow moss and has much the general appearance of an extensive fieldwork planned for defense by some *antediluvian*[17] *Carmontaigne* [Cormontaigne].[18] I should have attempted a plan of it but being informed it was strongly garrisoned by rattlesnakes, a reptile I have an intense aversion for, & finding no one of the party caring about a closer examination we proceeded to encamp some half a mile farther on in the bed of a lagoon where we fortunately discovered a small hole of water. Here taking up our Quarters beneath the shade of a huge Sycamore, bitten by mosquitoes & fatigued with our long thirty five or six miles we worried out the night in making commendable endeavors to sleep but generally with signal lack of success. The country passed over during the day from the lagoon, was entirely barren. A dry argillaceous[19] soil completely burrowed by squirrels and other ground animals, & with but a little of the poorer kinds of grass affords little satisfaction to the traveler & none to the settler for agricultural purposes. I should think as far as my observation extends that the valley of the San Joaquin excepting in certain places along the banks of the river is entirely unfitted for agriculture. At any rate as yet no one has cared to attempt it, among the earlier settlers of the country. On Sunday we left our encampment about 7 A.M. Capt. Halleck and myself rode

[17]Antediluvian refers to the time before the biblical flood or, in this case, to someone fearful of such a flood.

[18]Louis de Cormontaigne was a mid-eighteenth-century French military engineer, obviously studied by U.S. Army topographical engineers.

[19]A claylike soil is argillaceous.

over to take another look at the Fort and while there the party started. Returning to the Camp ground we found them absent & the ground being undulating we were unable to see them. We however after some slight examination struck their trail and overtook them after a hard trot of three or four miles. Our path lay in a N.N.E. direction over the valley of the San Joaquin here some twenty miles in width. We observed in many places, a rich argillaceous soil covered with the usual grasses, also Tansy Penny Royal, Milkweed, Sun Flowers, & several clusters of wild vines bearing gourds which when ripe are of the size of an apple.[20] I am not aware of the Botanical appellations of the above named plants but I presume that the majority will understand their character as readily from their common as their scientific description.

Proceeding onward for some sixteen miles, and observing several flocks of antelope in the distance after chasing four bewildered elk for a mile or two we came to a point opposite Livermore's pass in the mountains (which is discernible from the river on a clear day) here we struck directly toward the San Joaquin whose boundary we had distinctly marked for some time by the green belt of willows on the Horizon.[21] We found it impossible to get nearer the river at the place we first struck it than about 100 yards owing to the swampy nature of the bank which for that distance was entirely covered with Tula [tule] or bullrush, a species of reed growing on the low ground of both vallies [*sic*], which is partially submerged in the rainy season.[22] We accordingly proceeded up the outskirts of this swamp until we arrived at a little ridge, covered with oak timber which leading to the dry bed of a slough on the bank, enabled us to pass the tula [*sic*],

[20]Pennyroyal refers to two plants in the mint family, *Lamiaceae.* It has been used as a folk remedy. The gourd mentioned could be one of many. See *The Jepson Manual: Vascular Plants of California,* ed. Bruce Baldwin, Douglas H. Goldman, David J. Keil, Robert Patterson, and Thomas J. Rosatti, 2nd ed. (Berkeley: University of California Press, 2012).

[21]The expedition is just north of Livermore, California, on the west side of the San Joaquin River. Livermore was named for an English sailor, Robert Livermore, who received a land grant at the site in 1839. While the town went through a few name changes, it was finally christened Livermore, for its founder, in 1869. See Gudde and Bright, *California Place Names,* 179.

[22]Tule is a tall bulrush (sometimes as tall as ten feet) that thrives in freshwater marshes. The word itself is a derivative of the Nāhuatl term that referred to the large marsh bulrushes that existed in and about Mexico City. For a description of the various species in California, see M. Kat Anderson, *Tending the Wild: Native American Knowledge and the Management of California's Natural Resources* (Berkeley: University of California Press, 2005), 27.

just above the junction of the lower mouth of the Merced River with the San Joaquin. The ferry was about one mile below. Finding an old boat which had been used for crossing the slough when over flown we dispatched Sergt Carson with McNab to bring up the ferry boat, the men working the ferry having deserted their post. They returned in about an hour with an old whale boat, in which having placed all our packs, saddles etc. we embarked, and polled it down the stream the animals being driven through the swamp to a point opposite the landing place, in the meanwhile. The old boat swamped just as we landed wetting thoroughly all of our baggage including the rations of some of the men which they had with much forethought placed for that purpose in the very bottom. Having removed everything from the boat & spread our damaged property upon the hot sand to dry we turned our attention to the animals which crossed safely with the exception of our largest & finest mule which becoming entangled in his 'riata' was unfortunately drowned. We had now marched twenty miles and considering the tired state of the animals, and the fact that the next crossing on the Tuolumne was but two miles distant where we would be obliged to unpack the animals & repack them on the other side, we decided to remain encamped at this place for the night.[23]

We accordingly retired early in the vicinity of the Ferry tent which we found unoccupied but exhibiting traces of the hasty departure of the Ferryman in the presence of a half finished bottle of liquor and box of sardines. Nor did we have occasion to wonder long at his precipitation for immediately after sundown we were surrounded by myriads of mosquitoes which penetrating the mosquitoe bars [a material structure, likely of wood, forming a secure enclosure] and attacking us without mercy caused us a great deal of manual exercise, an entire loss of sleep and a strong disposition to profanity during the night.[24]

The astonishing numbers of these execrable insects, covering the banks of the San Joaquin and the low portions of its tributaries almost exceed belief, & must I should suppose prevent anyone from settling

[23]Tuolumne River, 150 miles in length, has its source in the Sierra Nevada. The grand canyon, or upper part of the river, includes the Hetch Hetchy valley, a part of Yosemite National Park where a controversial dam was built between 1913 and 1923, making the lower river valley mostly dry today.

[24]The "ferry business" was common in America, some remaining in one place for several years, and others disappearing in a season. It is impossible to identify the owners of the deserted "ferry tent."

upon or attempting to cultivate the soil in this vicinity, some of which is the most arable in the country. I was informed that a settlement was commenced and several log houses erected at the junction of the Stanislaus River & the San Joaquin and a portion of land put under cultivation promising excellent returns, but the mosquitoe season (about nine months in the year) coming on, the settlers were driven from their homes by these attacks and abandoned their improvement in despair.[25] To a person accustomed to the mosquitoes in the United States, upon the Mississippi and Ohio rivers it might appear absurd to advance the opinion that they could even prove a serious obstacle to the settlement of an advantageous business locality, but one night on the San Joaquin River, would probably remove any such impression very effectually from his mind. Our animals suffered very much as well as ourselves running about continually, rolling, & eating but little. We started as early as possible on the 9th to avoid the mosquitoes which continued their attack until long after Sunrise. Leaving about six o'clock we marched about N.E. for two miles. The San Joaquin and Tuolumne Rivers being distinctly marked by the bank growth of willows and tula [tule], when we arrived at the Ferry on the Tuolumne.

Here we were crossed in a whale boat in about two hours at an expense of $30, to *Tuolumne City,* which consists of an extensive sand bank and plain in its rear adorned by three or four tents occupied by a Mr. Piles & his family consisting of his son, with his wife and some children, a Mr. McDowall with a wife dangerously sick with fever and ague [malaria], and the ferry man.[26] The sand was blowing in clouds which

---

[25]Charles Maria Weber founded Stockton, California, in 1847. The original effort obviously proved untenable, and the community seemed to be abandoned when Derby's party passed by. Weber had earlier married a Mexican woman and acquired the grant of eleven leagues on the east side of the San Joaquin River, which became El Campo de los Franceses (French Camp). See Gudde, *Place Names in California,* 321; Raymond W. Hillman and Leonard A. Covello, *Cities and Towns of San Joaquin County since 1847* (Fresno, Calif.: Panorama West Books, 1985), 131–32. For a biography of Weber, see George P. Hammond, *The Weber Era in Stockton History* (Berkeley: University of California Press, 1982), 27–73, 90–98.

[26]Tuolumne City never became the metropolis that its founder Herbert Reichhold envisioned. Today it is the headquarters of the Tuolumne Band of the Me-wuk (sometimes spelled Miwok) Indian Tribe. Note the reference again to "fever and ague," or malaria. A malaria epidemic began apparently in 1846, at the outset of the Mexican War, and continued into the early 1850s. See Gary Clayton Anderson, *Ethnic Cleansing and the Indian: The Crime That Should Haunt America* (Norman: University of Oklahoma Press, 2014), chapter 10.

rendered it unpleasant without the tents. The mosquitoes were excessively thick and ravenous which rendered it more unpleasant within [the tents], the woman, Mrs. Piles, said the nights were too warm to use bars and the mosquitoes were too torturing to go without. They also looked miserably, and judging from observation I have no doubt their appearance did no more than justice to their feelings. We found here a "*cavallada*" [or] collection of horses, in charge of Mr. Piles from which Major Canby purchased a horse for $150 to replace that given out, which we left behind here it being unable to proceed further. We now continued our march over the prairie between the Tuolumne and the Stanislaus, passing a large drove of wild horses and seeing many elk and antelope at a distance, we came to a slough of the Stanislaus, where we found encamped a little off the road two men from Mazatlan [Mexico] with three servants, on their way to the mines. They had no other information to impart but that the mosquitoes were "*mucho brava*" in which opinion we entirely coincided. [We] Continued our march N of E for 5 miles on the bank of the Stanislaus, which we found a beautiful stream extremely rapid with high precipitous banks, covered with fine timber, oak, willow, alder etc. Here we found an American & Irishman about to encamp on their return to *Pueblo de San Jose* from Sullivan's diggings, but on finding that we thought of encamping in their vicinity they with much discretion repacked their animals and continued their march.[27] We returned about ½ a mile, on our route and encamped beneath a large oak, within a few hundred yards of the river where we were very little molested by the mosquitoes. A party of six Sonorans passed us from the mines in the course of the evening.[28]

We started on the 10th at seven A.M. and keeping the bank of the Stanislaus through an arid sandy soil for about six miles arrived at Taylor's Ferry immediately opposite [at] which Maj Miller with two companies of the 2nd regt. of Infantry was encamped.[29] I observed on

---

[27]Sullivan's Diggings was a rich gold discovery made by John Sullivan in 1848 on Sullivan's Creek, Tuolumne County. Gudde and Bright, *California Place Names*, 324.

[28]The term *Sonoran* seems to be used for nearly everyone migrating north from Mexico.

[29]Taylor's Ferry, operated by Nelson Taylor in 1849, was on the Stanislaus River, near modern Oakdale, California. Major Albert S. Miller, 2nd Infantry, obviously had just selected the camp, which he called Camp Stanislaus. The site, as Derby noted, was not ideal, and "Fort" Miller was later (likely in 1850) moved to the north banks of the San Joaquin River, where the state maps locate it today.

the route this morning large quantities of milk weed and wild gourd, & a vine called the California Pumpkin, which has small pumpkins the size of oranges hanging upon it in profusion.[30] We encamped the animals about one mile above the [Taylor's] Ferry & the General [Riley] & Staff crossed to Maj Miller's encampment by the boat which is a fine conveyance large enough to cross two wagons at once, & is drawn from one side to the other by ropes. We found Maj Miller's encampment pleasantly situated on the bluff, on smooth level ground covered with a fine grove of holly oaks and possessing withal the inestimable advantage, which we were well able to appreciate of having no mosquitoes. It has as a permanent site, however the disadvantage of becoming extremely muddy during the winter or rainy season, though the same may be [argued] I am informed of every otherwise eligible position in this part of the country. We remained at Maj Millers until the evening of the 11th when re-crossing the Ferry we proceeded to our encampment, a distance of one mile nearly East in a grove of fine oak timber. We left our encampment between 7 & 8 A.M. on the 12th having several new mules to be packed which we have obtained from Maj Miller, some of our original five suffering much from fatigue & sore backs produced by the badly constructed pack saddles with which we were furnished. We marched in a general N.E. direction keeping the Stanislaus in sight for some time on our left. This stream presented a beautiful appearance, its water extremely rapid but as clear as crystal, and its banks lined with willows, alder, mountain ash & other beautiful trees. It falls in some places over four feet in a mile, & being at this time swollen by the recent melting of the snow had quite a turbulent and impetuous character as it rushed over the large rocks & pebbles with which its bed is covered. The country now began to assume a different character. Hills replaced the level prairies over which we have been moving and the timber consisting of Holly, ~~Holly~~ Post & white oak, Ash, etc. became much more abundant. The ground ascended sensibly in every mile and before us stretched away to the right & left the spurs of the great Sierra Nevada whose white capped peaks were plainly discernible above their blue and indistinct outline of the

[30]The California pumpkin was likely *Cucurbita*, a gourdlike squash. See Baldwin et al., *The Jepson Manual*, 678.

range at a distance of 150 or 200 miles. Our road lay at first through a grove of the different species of oak. At a distance of six miles we came upon a tent occupied as an inn, when we procured fresh beef at 25 cents per lb. and where they asked three dollars per bottle for ale. But we did not buy any of it.

The road from this point became much more rough & covered here & there with broken lumps of lava & trap rock. Eight miles further on we came to the 'Mountain Inn' a widely pitched tent occupied or kept rather by a single enterprising youth who with a bag of hard bread, one of coffee and three bottles of whiskey had determined to try his fortune in the public line. We met & passed many parties on their way from & to the diggings, & observed that either from policy or fear those returning invariably gave bad accounts of their success. On approaching or entering rather the region in which gold is found the face of the country undergoes an entire change. The color of the soil becomes suddenly a yellowish red, produced by the presence of ochre and oxide of iron from disintegrated rock, and the ground is thickly strewn with fractured pieces of the white porous quartz which is generally indicative of the presence of gold. Schistose Slate is also abundant cropping out in singularly shaped strata which standing in rows are eminently suggestive of graveyards, & may serve as hints to the inexperienced gold hunter of the perils & dangers of the region he is entering.[31] Gold is frequently found in thin plates between the layers of this rock, as well as injected in the crevices of the porous quartz. Everything gives evidence of strong volcanic agency. The gold bearing strata is supposed to be from thirty to forty miles in width and if exposed at the time the Sierra Nevada mountains were upheaved [*sic*] it is to be supposed not only from theory but analogy that a similar stratum exists upon the eastern side of those mountains occupying an analogous position with reference to the surrounding rocks with this. We kept on up & down the hills through this wild country keeping generally about 15° N. of our first bearing until we arrived at a large plain, lying between the tributaries of the Stanislaus & Tuolumne,

[31]*Schistose* is a term denoting stages of flakes of rock, sometimes resembling mica. Here Derby refers to the foliation that occurs with slate.

surrounded by oak timber, & covered with a very good growth of fine grass. Here we encamped by a beautiful stream of water running from a spring in the hills, having marched 32 miles and being but one mile from 'Wood's diggings' the first mines upon the Tuolumne.[32] Before reaching our encampment we had passed over a plain stream with pieces of lava on each side of which rose singularly shaped hills, the summits of which cropped out with trap rock much in the shape of the natural fortification noticed above & giving the idea of a high stone wall bordering the summit of each. This day was excessively warm, and we observed a great change in the temperatures continually after the heat in the mining district being intense while the valleys & plains below, from the constant current of air, are comparatively cool. We left our encampment at 8 A.M. on the 13th and riding nearly East soon came to Wood's Creek, a small tributary of the Tuolumne, nearly dry at this season but with holes of water standing in the bed of the stream amply sufficient for the miners purposes. Here we observed for the first time the process of gold washing[.] Some fifty men of all nations languages & color were busily engaged, in the bed of the creek, usually in small parties or gangs of from three to five individuals. The ~~instrument~~ machine used by all was the common Rocker. This is made like a common child's cradle, open however at one end & slightly inclined, with two or three cleats, two inches high nailed across the bottom on the inside: at the upper extremity is placed a coarse sieve beneath which an apron placed at an angle of 45° and reaching nearly to the bottom of the trough is nailed. The stones are first thrown from the bed of the creek, the earth exposed is then loosened with a small pick, or a shovel, & thrown upon the sieve, one man then rocks the machine from side to side by an attached handle, while another continually pours on water. The pebbles left in the sieve are thrown away & the more dense particles of earth including the magnetic black sand which invariably accompanies the gold collect against the cleats.

---

[32]Wood's Diggings was named for Benjamin Wood, who along with James Savage, employed Miwok Indians to open the diggings. See Daniel S. Woods, *Sixteen Months at the Gold Diggings* (New York: Hayes & Brothers, 1851), and Albert L. Hurtado, *Indian Survival on the California Frontier* (New Haven: Yale University Press, 1988), 112–16.

Some five or six bushels of earth being thrown on & the worthless portion washed away, the residue is collected by pouring water into the trough and opening plugged holes behind each cleat, a large bin or wooden basin being held beneath to receive it. One person then takes the basin to the water's edge and continually dipping it in the stream gives it a rotatory [*sic*] motion which gradually washes away the remaining earth and most of the black sand, leaving the gold nearly pure. When dry the sand remaining is blown away by the mouth. This I observed is the usual method of collecting the gold & by working steadily the operators average from one half to two ounces a day each. The mines procure coarse fare from the trading tents, (one or two of which were in operation at this locality) at enormous prices, but still unless worthless and extravagant may save enough to repay them handsomely for their labor. Leaving Wood's we proceeded about six miles to a collection of tents upon another creek emptying into the Tuolumne called Jamestown, where were two or three hundred persons at work in the same way, but generally with rather better results, two ounces per diem being a very common yield for one individual.[33] Here we found many trading tents, most of them bar rooms & groceries, the proprietors of which unless their own best customers, do rather better in accumulating the gold than the actual collectors. From Jamestown we went to the Sonoran camp on the Tuolumne, a large collection of tents laid off regularly in streets & containing some 1500 or 2000 inhabitants principally Mexicans. This we found an excessively busy encampment, drinking & gambling being the steady occupation, & mining the occasional amusement of a large portion of the citizens. An *alcalde* was in the exercise of his functions & at the time we arrived eighty two trading tents of which fifty four were either bar rooms or

[33]Jamestown, Tuolumne County, was a gold rush town opened by Colonel George James, which today is a California historical landmark. It was located just above Wood's Diggings, Derby making the distance a mere eight miles. Gudde notes that the early settlement was for a time broken up when disputes occurred with the Mexican occupants, who, according to Derby, seem to outnumber Americans. See Gudde and Bright, *California Place Names,* 155. Jamestown has been described as a district eight miles in length from northeast to southwest. See William B. Clark, *Gold Districts of California,* Bulletin no. 193 (Sacramento: California Division of Mines and Geology, 1980), 17, 77.

gambling establishments, were in the full tide of successful operation.[34] I observed large enclosures surrounded by cotton cloth, and erected in a circular form some hundred feet in diameter where on Sunday the Sonorans enjoy their favorite national pastime of bullfighting.

We encamped for the night on a little creek, emptying into the Tuolumne and about three miles from the Sonoran camp & on the 14th visited in succession Sullivan's and Curtis's Creeks on the former of which the miners are succeeding particularly well and which has the reputation of being the richest "placers" yet discovered.[35] I observed that at both these settlements pumps were used connected by wooden troughs with the rockers, which last were generally somewhat larger, but on the same principle with those used on Woods Creek. Returning we passed again through Jamestown having made a circuit of about twelve or fifteen miles, and encamped on the same spot as the night previous. The prices of everything at all these "placers" were enormously high, forage particularly, one dollar being the price for a small bundle of grass six inches in diameter. On the 15th we started at 9 A.M. rode W.N.W. for four or five miles where we crossed the ferry on the Stanislaus in a small boat & swimming our horses over. The Ferry is about one mile below the locality known as the "French diggings" and is established at the base of a high range of hills running parallel to the river & covered with the typical rocks, porous quartz, Schistose Slate etc.[36] Future collectors of curious geological specimens will undoubtedly be gratified some hundreds of years hence at the immense tracts of fossil bottles of antique shapes (all emptied however) which they will cross in explaining this region of the country. Having packed our animals and saddled up, we proceeded on in a S.W. direction for some 11 miles, crossing a continued succession of high hills and passing Carson's Creek which like that of Angel [Angel's Creek] some six miles farther has been extremely productive but is now entirely

---

[34] The presence of an alcalde suggests a form of city government, since an alcalde was in essence a mayor. The diggings were generally called "Santiago."

[35] For early mining at Sullivan's and Curtis's Creeks, see J. D. Munro-Fraser, *History of Santa Clara County, California: Including Its Geography, Topography, Climatology, etc.* (San Francisco: Alleg, Bowen, 1881), 698–99.

[36] This region apparently produced little gold and disappeared early.

deserted.[37] The first three miles of this day's journey were on the bank of the Stanislaus river which we found to be falling rapidly, & in the vicinity of which were collected many parties of miners, principally Sonorans, waiting for the water to subside sufficiently to allow working the exposed sand bars which are believed to be extremely rich. The hills & bank of the river were thickly studded with Holly, Beach, White & Post Oak & two or three different kinds of Pine. I noticed particularly the common white & the yellow or hard pine. The Pine is found throughout the gold region, and appears like the quartz rock to be an indication of the presence of that mineral.

Near Angel's Creek, on the right of the trail we passed a spring of highly sulphuretted [*sic*] water which is in great repute among the Indians for its medicinal properties, it was entirely surrounded by holes made by them for bathing in. We stopped at several hopeful looking places to search for gold, and after laboring for an hour in a cliff of Schistose Slate my exertions were crowned by the discovery of an undoubted specimen nearly as large as a pin's head. The other gentlemen were I believe unsuccessful.

Game appeared to be scarce among these hills. We observed two large hares during the day, but they were excessively wild. We encamped at night at the foot of a range of hills about 18 miles S.S.W. of the Sonoran camp where we found excellent water and good grass of which our animals were sorely in need. On the 16th we marched at 8 A.M. for 22 miles in a direction varying between S. & S.W. and encamped on the Calaveras (or river of skulls) a small, crooked stream

---

[37]Sergeant James Carson had discovered gold at Caron's Creek in late fall 1848, then abandoned the mine. He was with the Derby party as guide, thus explaining Derby's remarks regarding early productivity. After abandoning the mine and leaving the army, Carson was elected to the state legislature, but he died in 1853 before assuming office. A Mexican named Pacheco, perhaps the same man as the owner of Rancho Ausaymas, rediscovered a much larger vein in 1850 at Carson's Creek. The mine later produced $3,000,000 in gold in a few years but then was shut down due to legal battles regarding ownership. Angel's Creek or Camp was started by Henry Angel in 1848. Rich surface ores were found in the 1850s, and the mine continued operation for the next sixty years, producing an estimated $30 million. The Jumping Frog Jubilee at Frogtown, in the heart of the region, is the subject of Mark Twain's story "The Celebrated Jumping Frog of Calaveras County." See Gudde and Bright, *California Place Names,* 11, 55; Clark, *Gold Mining Districts of California*, 25.

emptying into the Sacramento River near its mouth.[38] The first four or five miles crossed the range of hills that we had ascended on leaving Miller's camp, & amidst which the "diggings" commenced. These hills vary in height from 6 [600] to 800 feet. About 80 miles E. we saw from the summit of the range we crossed a very high peak from the summit of which Sergt Carson informed me a view of the whole country may be had.

Descending gradually we entered again upon the San Joaquin plains much oppressed by heat and thirst, all the springs that we passed being excessively bad water, entirely unfit to drink. One horse gave out entirely & we were obliged to leave him behind. We saw today as heretofore many "coyotes" a hybrid animal between the fox and wolf, which combines all the bad qualities of both with none of their good—supposing them to possess any—these detestable animals are very numerous though the whole of upper California. We met in crossing the plain a train of forty or fifty packmules [*sic*] on their way to the mines with goods from Stockton. On leaving the gold region we observed the change in the color of the earth very apparent, the soil on the plain clayey. Before striking the Calaveras we crossed a range of low Sand hills, direction S.E. & N.W. We found on the river bank a profusion of the wild grape vines covered with unripe fruit. The river was extremely low, no mines have as yet been discovered in its bed. The grass at our encampment was excellent nearly breast high affording luxurious feed for the animals. On the 17th we marched at 7 A.M. for 14 miles encamping on the bank of the Moculumme [Mokelumne River], which stream running from the hills is variously stated to empty into the Sacramento & the San Joaquin.[39] Carson states that it discharges itself into the estuary, or swamp, between the two at the head of Suisun Bay & his information being generally reliable I have adopted his idea

---

[38]The Calaveras River, fifty-two miles in length, actually joined the San Joaquin River west of Stockton. Its name came from the discovery of a vast array of skulls, probably of Miwok Native Americans killed by a Spanish expedition.

[39]The Mokelumne River got its name from the Miwok Indians, the band's name being "people of Mokel." See David L. Durham, *California's Geographical Names: A Gazetteer of Historic and Modern Names of the State* (Clovis, Calif.: Word Dance Press, 1998), 804. See also Kroeber, *California Place Names of Indian Origin* (Berkeley: University of California Press, 1916), 48.

on the subject. On crossing the Calaveras a mile below our encampment we passed a tent kept by a soldier on furlough who treated us to some very cold water from an excellent spring on the bank. He had charge of a large corral of horses belonging to persons working in the mines. The land in this vicinity had been subjected to cultivation by the Indians, whose deserted Rancheria near by built of boughs of trees, long since withered, resembled a collection of vast birds' nests. We crossed the plain which appeared to be excellent land for cultivation, and ascending some low hills met several Indians all well clothed, but with an evidently uncomfortable appearance, looking as though they would much prefer their pristine state of nudity. They are all excessively ugly particularly the females & upon the whole somewhat resemble the Winnebago's on the Upper Mississippi. A little farther on we passed their Rancheria where were some forty or fifty of them. Observed one squaw adorned with a brimstone colored petticoat which gaudy article of dress she had tied tight around her neck to the utter elimination of her arms, presenting the appearance of a magnified pin cushion. Some of the squaws carried on their backs large tunnel shaped willow baskets, supported by a back strap passing round the forehead. In these baskets I am informed they frequently carry provisions etc. of the weight of 100 pounds.[40] Passing the hills we came out on a rich plain of fine land, crossing which we passed through a heavy growth of oak timber. We kept a little to the West to avoid crossing a thicket of "chisinel" (species of fir) where we saw many horses which are much larger than those in the United States and run with great rapidity. The ground here was thickly covered with small crystals of quartz, & with garnets in profusion, but I observed no very fine specimens. The rattlesnakes are extremely thick in this region of the country. We shot one in the hollow of an oak tree, having seven rattles, and the middle of the body two inches in diameter. A singularly shaped hill served as a point of direction for the Mocullumme [Mokelumne], where we

[40]The Miwok Indians, whom Derby describes here, dominated the Sierra Nevada range as well as the Central Valley below it, from the Fresno River in the south to the Cosumnes River in the north. See Alfred L. Kroeber, *Handbook of the Indians of California* (Washington, D.C.: Bulletin 78, Bureau of American Ethnology, Smithsonian Institution, 1925; reprint, New York: Dover, 1976), 442–61.

arrived about one o'clock & encamped for the night. We observed on the road milkweed, penny royal, southern wood, & the rattle snake weed, said to be an infallible remedy for the bite. Fine grass in this vicinity. Probably among the richest gold washings in the country may be panned in the bed of the Mocullumme.

On the 18th marching at 20 minutes past 7 A.M. we proceeded on the bank of the river 12 miles to McKenzies' Rancho, situated on a little hill and including 5 or 6 miles of most excellent land, the only draw-back being the numbers of mosquitoes which are very troublesome.[41] The Rancho is now the property of a Mr. Laird, who with his family received us very hospitably. The day being very warm, & the animals suffering much from heat & fatigue the General concluded to encamp here for the night which we accordingly did, on a hill in the rear of McKenzies House where we suffered much from the mosquitoes all night. Laird has here an excellent farm, not well stocked at present however. He has plenty of milk & manufactures excellent butter and cheese. There were two or three men hard at work manufacturing gold washers. They informed us that the river was extremely rich some 60 miles above, but the miners generally dislike to go there on account of the Indians who are represented as extremely troublesome. Leaving our encampment at 7 A.M. on the 19th we crossed the Ferry in a small "dug out" swimming the animals over, which operation detaining us some two hours amid swarms of venomous mosquitoes was not of the most agreeable kind. Our road lay at first through the oak timber upon the river bank (all the rivers in this region of the country have a heavy bank growth of timber by which their courses may be distinctly marked to some distance from the plains) then crossing a dry arid plain for some 5 miles we crossed the Rio Seco or Dry creek which running nearly parallel to the other mountain streams, loses itself in the marshy ground at the head of the Suzun [Suisun] Bay.[42] As usual at

[41]McKenzie's Rancho does not appear in any other source, and he may be referring to George McKinstry, an early employee and later adviser to John Sutter. McKinstry went into the mining business in 1848, apparently supplying goods and foods. See Albert L. Hurtado, *John Sutter: A Life on the North American Frontier* (Norman: University of Oklahoma Press, 2006), 237, 245, 296.

[42]Rio Seco translates as Dry Creek.

this season of the year, we found but the dry bed of its stream, but with some four standing holes of water, the presence of which was marked by the willow in the vicinity. No signs of mineral region. Some fir, beach [*sic*] and ash trees on the bank and larger quantities of post oak.

The next 15 miles we passed through large quantities of wild oats which the animals eat with avidity and arrived at Shaddon's Rancho,[43] a large & well cultivated farm on the Southern bank of the Cosumne [Cosumnes] River. This is a small rapid stream flowing through a rich country & discharging itself into the Estuary at the confluence of the Sacramento & San Joaquin. It is supposed by some persons to be a branch of the Moculumme [Mokelumne] but I could obtain no satisfactory evidence that such was the case. We forded it, a little above Mr. Shaddon's hut and encamped beneath a huge oak on the Northern bank. Leaving the Cosumne [Cosumnes] at about the usual hour we marched for 15 miles over a dry sandy plain, with intervals of dry wild oats and grass, on which herds of cattle & *caballadas* of horses were grazing, Mt. Diabolo [spelled Diablo today] was plainly visible in the S.W. at about 40 miles distance and after proceeding about 10 miles we could plainly discern the line of trees marking the direction of the Sacramento River and farther in the distance that of the American River. Sutter's Fort is situated on the rising ground just above the confluence of these streams.[44] Arriving within two or three miles of the Fort, whose adobe walls and tiled roofs, surmounted by the American Flag looked very prettily in the distance, we turned to the west and leaving the direct road to the Fort struck the direction of Sutterville, a position on the Sacramento where a large city has been projected, but which at present is occupied by a few hastily constructed shanty's, the inhabitants of which are call "Suttervillains" whether appropriately or not I

[43]Thomas J. Shaddon supposedly purchased "a League" of land on the river from "Chief Heleno" of the Mokelumne band of Indians. He then sold it in 1850 for $5,000, whereupon another claimant appeared and challenged the deed. See the long discussion of the court case that followed in 1864 in Charles A. Tuttle, *Report of Cases Determined in the Supreme Court of the State of California* (Sacramento: O. M. Clayes, 1865), 25: 127–47.

[44]Mount Diablo rises some 3,849 feet in Contra Costa County. From the top, most of the San Francisco Bay area is visible. Today, the peak is a state park. Sutter's Fort on the American River was built by John Sutter in 1839 and became the way station for early immigrants into California. See Hurtado, *John Sutter.*

am unable to say.[45] The portion of the "city" on the river is a horrible sand bank infested with ravenous tribes of mosquitoes; immediately in rear of which is a swamp some half mile in width beyond the swamp it extends into the plain where I believe it is not impossible to live.

Maj Kingsbury we found encamped below this place with one company of the 2nd Infantry in the midst of a well attended convention of mosquitoes, and we remained here for the night.[46]

On the 21st we rode to "Sacramento City," the landing place for vessels at the confluence of the Sacramento & American rivers, which we found a "cotton city" [meaning tent city] of spacious dimensions, containing 2 [2,000] or 3000 inhabitants. Some 12 or 15 vessels were lying at the bank & the place wore an air of business and enterprise quite refreshing to behold (There is [word "omit" again added][47] but one objection apparently to this becoming a large commercial place & that is the trifling disadvantage of being under water for four months in the year.[48] This might I suppose be obviated by the construction of a levee upon the bank.) We remained at Major Kingsbury's encampment until the evening of the 22nd when we moved to the rear of Sutter's Fort where we encamped beneath some fine oak trees in the vicinity of a small lagoon, in preparation for an early start the ensuing morning. Our march on the 23rd was painful the heat being intense. The sand rose in clouds around us the soil on this plain being much more siliceous [containing silica] than on those farther South.

---

45 Suspecting the possible publication of the journal, someone, likely on General Riley's staff, entered the word "Omit" just in front of the phrase "the inhabitants," and continued to use the term all the way to the end of the paragraph, or where it reads "not impossible to live." The tone of Derby's report, witty and sarcastic at times, may be the reason why it was not sent east for publication in the serial set.

46 Major Julius J. B. Kingsbury commanded a battalion of the 2nd Infantry and had received orders to locate a military post north of Sutter's Fort.

47 The part slated for omission included two sentences ending with the phrase "levee upon the bank."

48 A stiff competition was underway between Sutterville, a site three miles south of the junction of the two rivers, and present-day Sacramento, the "cotton" city that Derby sarcastically noted would be prone to flooding. The flooding did occur in the two years that followed, and Sutterville supporters thought it would benefit their city. But the founders of Sacramento built dikes and saved the town, which doomed Sutterville. The competition is well documented in Hurtado, *John Sutter*, 245–69.

The soil on the sides of the road had however the appearance of being extremely fertile. We struck the American Fork near Sinclare's [Sinclair's] Ferry at about 3 miles distance from the Fort and proceeded on this bank of the beautiful stream for some seven or eight miles farther when we turned to the left from the road and encamped in a rich bottom near General [Persifor F.] Smith's encampment among a fine grove of heavy oak timber.[49] We passed during the day over several recently cultivated wheat fields which I am informed have been extremely productive. Trees, oak, holly & willow, Grass, coarse but nutritious, what is called meadow hay in New England I observed to be very plentiful, no mosquitoes or but few, though sand flies and pismires [archaic for ants] are abundant and render sleeping on the ground very uncomfortable.

Maj Canby and Captain Westcott hearing in the evening of the arrival of Capt Day & his command of the 2nd Infantry at "Sacramento City" concluded to return to that place and rejoin us the following day.[50] Accordingly we started without them on the 24th in a N.E. direction for Mormon Island which we reached about 12 n. marching about 12 miles.[51] This place is on the S. Fork of the American River about two miles from its junction with the N. Fork. The current above the Island is very rapid but by the energy of the party working there consisting of 28 members of an incorporated company they had nearly completed a dam and canal by which the bed of the stream on the N.W. side of the Island will be exposed and judging from the amount of gold already collected from this Island will amply repay them for the months of toil expended. There were several parties at work on the

[49]John Sinclair, a Scotsman who came to California in 1839, was closely associated with John Sutter. See Marguerite Eyer Wilbur, ed., *The Adventures of Heinrich Lienhard* (Los Angeles: The Calafia Society, 1941), 30. Sinclair was elected alcalde of Sutter's settlement, New Helvetia, in 1847. See Hurtado, *John Sutter*, 203.

[50]Captain George C. Wescott and Captain Hannibal Day were both with the 2nd Infantry. Day later took command of Camp Far West, located on Bear Creek. Captain Justus McKinstry, the quartermaster, later left an account of operations at Camp Far West, found in John K. Driscoll, *Rogue: A Biography of Civil War General Justus McKinstry* (Jefferson, N.C.: McFarland & Company, 2006), 80–89.

[51]Mormon Island was fifteen miles west of Sutter's Mill (Coloma), a mine opened by two Mormons, Wilford Hudson and Sidney Willes. See Hurtado, *John Sutter*, 203. For the complete story of Mormons in the Gold Rush, see Kenneth N. Owens, *Gold Rush Saints: California Mormons and the Great Rush for Riches* (Spokane, Wash.: Arthur H. Clark, 2004).

Island, whose tents were pitched on the bank of the river, and the high bluffs in the vicinity. [We] Observed here much fine timber, pine & oak, and a fall of water on the stream amply sufficient for working a saw mill. A Mr. Brinsmade had a party of men working with a large Rocker, which having mercury placed behind several copper cleats in its trough, amalgamated all the gold washed over its surface & the amalgam being afterward roasted on a crucible produced he informed me very profitable results.[52] The machine is so heavy however & the cost of transportation so high that it will never probably be introduced to any great extent in working the mines. The number of miners here I estimated at about 250. We left the Island at 5 P.M. on the 25th Maj Canby having rejoined us in the morning and moved some seven miles through the hills (which were covered with fine timber) to the Green Spring.[53] Here we found a beautiful interval, covered with grass, and watered from an exceedingly cool and refreshing spring, near which two negroes had established a trading house, which is also a stopping place for the newly established stage line. We encamped in the vicinity under a fine oak tree & on the 26th leaving at 8 A.M. arrived at Sutter's mill or Colluma [Coloma] after passing through a hilly country well wooded with pine and oak. Colluma is settled by some 3 [300] or 400 enterprising people principally traders & mechanics.[54] The buildings are almost all of wood and substantially erected. There is a fine saw mill at work, furnishing 1500 feet of timber per diem for which the owners ask and receive the enormous price of $500 per thousand.[55] We observed here the spot in a ravine near the principal street where the gold was first discovered.

---

[52]Peter Allen Brinsmade (1804–59), a graduate of Bowdoin College, first attempted to build a commercial sugar plantation in Hawaii. He sailed for California in 1848, worked as a newspaper editor, then came back to try mining. See Paul T. Burlin, "Peter Allen Brinsmade and the Tragic Pursuit of Pious Capitalism," in *Imperial Maine and Hawai'i: Interpretive Essays in the History of Nineteenth-Century American Expansion* (Lanham, Md.: Rowan & Littlefield, 2008), 21–56.

[53]Green Spring is located eight miles southwest of Coloma, California.

[54]Coloma, named for a Maidu Indian village nearby, was a settlement that developed near Sutter's Mill. Sutter spelled it Culloma, explaining Derby's Colluma. The present spelling came into use shortly after Derby's visit. See Gudde and Bright, *California Place Names,* 71; Clark, *Gold Districts of California,* 39.

[55]For whatever reason, Derby fails to mention John Sutter, the owner of the mill and the most recognizable person in the region.

Colluma with its mills, its favorable location for trade with the mining population and above all the enterprising character of its inhabitants bids fair in my opinion to become one of the principal cities in the upper country. The General who had been suffering for some days from ill health and fatigue, finding himself too much indisposed to continue his route as intended into the mining districts on the tributaries of the Sacramento, concluded to return from this place, and declining a public dinner tendered him by the citizens of Colluma started on the morning of the 27th for the Green Spring where we again encamped. The Kannaka Camp containing some 600 Sandwich Islanders, is situated on the S. Fork some 6 miles north of this place, but I was informed that owing to the idleness & extravagances of these people, though in a favorable locality, they are not working to much advantage.[56]

On the 28th we marched some 21 or 2 [22] miles in a southerly direction over a fine rolling prairie passing three trading tents at intervals of four or five miles upon the road & crossing the Cosumne [river] again below Daly's Mill two or three miles from which we encamped. We past [*sic*] on the road many emigrants from Missouri with wagons who stated that over 5000 persons with 1500 wagons were already on this side of the mountains on their way to the Gold region, many intending to dig at Mormon Island, others on the N. Fork etc. The hills south of the Green Spring are covered with low brush wood, the first of the kind I had noticed. The soil generally over which we passed was extremely fertile, and at Daly's Rancho, where are several buildings and an old mill, a fine field of wheat had just been reaped, apparently very productive. Leaving our encampment early on the 27th we moved over the same plain which we had crossed below in making Shaddon's Rancho, & passing the Rio Seco we struck the road to McKenzie's Rancho which we reached about 12 m.[57] We found McKenzie's family in deep affliction having lost their youngest child within a few days very suddenly from something alarmingly like Cholera & the wife of a Mr. Dunleavy

[56]Kanaka, a reference mainly to Native Hawaiians, was used to identify various miners who came from the islands. Green Springs, no longer in existence, was located eight miles southwest of Coloma in the neighborhood of present-day Rescue. See Durham, *California's Geographical Names*, 404, 545.

[57]For Shaddon's Rancho, see footnote 43.

was at that time lying in the house at the point of death. The whole region of country upon these rivers is now becoming unhealthy, fever and ague being the prevalent diseases, from which many die, owing in great measure to the want of proper medical attendance & comforts & conveniences of life.[58] Marching on for 8 or 10 miles we struck the Calaveras upon the Rancho of Dr. Isabel, where we observed about 100 head of horses and cattle feeding upon the excellent grass in the vicinity. The river was nearly dry, but some pools of water were left standing in its bed amply sufficient to supply our wants. Leaving at an early hour on the 30th and riding about 8 miles S.W. through Oak openings, we came suddenly upon the masts of shipping towering above the trees, and crossing a small belt of tula formed across the head of the slough on which the place is located we found ourselves in Stockton.

This place is at the head of a slough communicating with the San Joaquin river, and from its proximity to the mines on the tributary streams, and the fact that vessels of large size can enter the slough and be close to the bank, will undoubtedly become the principal commercial city in this part of the country. At the time we visited it, we found but one wooden building (a large shop belonging to Mr. Weohre [this may be Charles Weber, the founder of Stockton]) the other tenements being large tents made by stitching cloth over heavy frames, but there were about a dozen vessels lying at the bank, barques, brigs, & schooners, most of which are metamorphosed into shops or groceries.[59] We encamped in the vicinity, and leaving on the 31st at 9 A.M. passed the French encampment at a distance of 4 miles S.E. when we found the Camp of Maj Graham, who was on his way to Stockton on private business and from thence marching W. of S. over a level prairie for some 12 miles struck the San Joaquin at the Ferry about 8 miles above the junction of the Stanislaus. Here we crossed in a large scow drawn by ropes from one side to the other [and] marching 5 miles farther came to a fine lagoon of excellent water where we encamped. On the 2nd of August leaving encampment at ½ past 6 A.M. we marched 33

[58]The malaria epidemic had continued throughout the summer and into the fall. Cholera also had become a killer, likely being brought in from Central America.

[59]For Weber, see footnote 25.

or 4 [34] miles nearly N.N.W. and encamped near a small lagoon 15 miles N.E. of Mt. Diabolo [Diablo] and in sight of Suzun [Suisun] Bay, the first part of our march over a wide & almost barren plain extending from the coast range to the river San Joaquin, where we saw immense herds of elk and large numbers of cattle; we found no water on our trail for the first 30 miles and suffered much in consequence. On arriving at the lagoon and encamping Sergeant Carson went out and killed an antelope, which was the first I had ever seen. It looked very like a species of wild goat, and its flesh proved delicious eating resembling much goat flesh in flavor.

On the 3d, leaving our encampment at an early hour we marched some three or four miles from the lagoon in the direction of the bay, when at the junction of the San Joaquin & Sacramento rivers we espied two or three slight frame buildings and a tent erected on small eminences of land running up from the tula swamp which lines the shore in every direction. [Here an editor added another "omit."] These buildings were occupied by two carpenters (Dupee & Smith) who on the promise of $14 per diem from Col Stevenson had voluntarily banished themselves from a habitable country to erect buildings on this spot, where I was informed a city had been projected, & sarcastically termed "New York of the Pacific."[60] Stopping for a few moments & enduring the bite of myriads of mosquitoes for that length of time, to condole with these unhappy men, I at length turned away and followed the remainder of the party who by this time had nearly crossed the spur of the Mt. Diabolo. I observed '*en passant*' [in passing] that there was no wood, or even a tree of any kind within three miles of the projected city [parenthesis for ending the "omit" edit appears here]. Crossing the spurs of Mt. Diabolo at a distance of 15 miles from its foot we came to Martinez, a point on the Straits of Carquiniz [Carquinez] opposite Benicia, the tents and buildings of which place are in full view on the other bank.[61] At 5 P.M. the General & staff crossed to Benicia, and

[60]Colonel L. D. Stevenson had commanded volunteers at Los Angeles in 1847 and then mustered out of service. Many of his letters during 1847 are found in NARG 393, LR, 10th Military Department. Derby was correct in asserting that New York would likely not survive.

[61]Don Ygnacio Martínez owned a large ranch on the south shore of the Bay of Carquinez Strait.

concluding on the following day to proceed by water to San Francisco, I received instructions to proceed with the men & animals to the vicinity of Santa Clara, there to remain until I could hear by express the General's intentions as to returning to Monterey by sea, or marching from San Francisco by the usual land route. I accordingly started with the party on the evening of the fourth of August & marched 15 miles S. by encamping under a tree near a small unoccupied building on the usual trail. On the 5th I marched 26 miles S.E. and encamped on the road, at a little creek which I believe empties into the Bay of San Francisco. We left at 7 A.M. & found the road very pleasant, a fine country well timbered & watered and excellent land for cultivation. We passed several fine ranchos with large dwellings situated pleasantly upon them & numerous herds of large cattle. On the 6th we arrived at Santa Clara when we encamped on the stage road at about ¼ of a mile from the village. The valleys of San Jose and Santa Clara lying on the Coast Range, are with Napa valley the finest land for agricultural purposes in this part of California. We passed in the valley of Santa Clara through large tracts of wild mustard the pods of which I examined and found the seed perfectly ripe & of excellent quality. All the land under cultivation does extremely well, particularly the wheat fields which are very productive. At the Mission de San Jose 15 miles N. of the Pacific we found an excellent orchard of pear & fig trees loaded with fruit judging from which I should suppose there would be no difficulty in cultivating any description of fruit with success in these valleys.[62] Having remained in camp near Santa Clara until Thursday Aug 9th without receiving any communication from the Commanding General, I concluded to leave for Monterey. Sergt Ferry being very sick I left him in charge of Mr. Bellomy at Santa Clara and engaged Dr. Cobb of San Jose, the only physician in the vicinity to attend him. Leaving our encampment on the 9th I arrived with the party at Monterey on Saturday Evening the 11th encamping near Murphy's rancho on the 9th and at our first encampment near San Juan on the 10th.

[62]Mission San José was founded in 1797 by Franciscans and declined rapidly after secularization in 1836.

I have the honor to enclose herewith a general sketch of the route, which, made without instruments of any kind, merely from personal observation & what information I could obtain from Sergt Carson and others familiar with the country, can not of course claim much merit for accuracy. The outline of the coast & bays is from an old Spanish map in the possession of the Department and is I believe generally accurate. That of the direction of the coast range and the more important rivers I assumed from my own observation. It will be seen that by the sketch, the Cosumne [Cosumnes River] empties into the Sacramento and the Rio Seco, Mokelumne, and Calaveras into the Estuary at the head of Suzun [Suisun] Bay. This is disputed by many persons who claim to have examined these rivers to their mouths & find them emptying into the River San Joaquin.[63] I do not know that this may not be the case, as far as personal examination extends but from the information I have received on the subject, which ought to be reliable, & from observing the general direction of the streams I am inclined to think my sketch in this respect more correct than those which have preceded it. If I had had the advantage of instruments, assistance, and an opportunity to examine the corrected map of the coast & country between this place and San Francisco[,] I should undoubtedly have presented an entirely different piece of work. As it is it will at least serve the purpose of defining the route actually passed over by the General which when a correct examination and map of the country is made, it may be easily transferred to the latter. For the above reasons I have not attempted to execute a more finished map than the sketch I have the honor of presenting.

With high respect
Your obdt servt
Geo H. Derby
Lt Top Engrs

[63]Derby had only a vague idea of the mouths of these rivers, likely getting the information from Sergeant Carson. With the exception of the Cosumnes River, which empties into the Mokelumne River, the others generally are tributaries of the San Joaquin River.

Bennett Riley's Assessment of Indian Affairs in California July–August 1849

General Riley learned much on his reconnaissance with Lieutenant Derby. He subsequently wrote two reports that derived from his trip into the interior, mostly concentrating on the need to control the California Indians adjacent to and located in the mining district. Since these reports shed some light on Derby's manuscript, they are included below.

Letter to Lieutenant Colonel W. G. Freeman

Head Quarters, 10. Mil. Dept.[64]
Monterey Cal. Sept.20. 1849

Colonel [W. G. Freeman]

I have the honor to transmit herewith a copy of a topographical sketch, by Lieut Derby, Topographical Engineers, of the country traversed by me on a recent reconnaissance of a portion of the San Joaquin and Sacramento vallies [*sic*] and to submit for the information of the Commanding General some remarks upon the character of the country lying between the Coast range and the Sierra Nevada and between the confluence of Feather River & the Sacramento on the north and the lake Buena Vista on the South. These remarks are based upon my own observations and upon information abstracted from the reports of Officers of my command made in accordance with instructions heretofore given by me.[65]

My original intentions, as indicated in my report to Division Head Quarters of June 2nd was to have proceeded by the pass of San Miguel to the lake country east of San Luis Obispo for the purpose of determining the proper position for the establishment of a military command to prevent the incursions of the Tulare Indians (recommended in my report of April 25th) and to give protection to the mining interests north of the lake.[66] Information of the hostile attitude assumed by the Indians

[64]Riley to Smith, September 20, 1849, NARG 393, LR, DP.

[65]Buena Vista Lake was at the south end of the Central Valley, fed by the waters of Kern River.

[66]San Miguel Pass started through the coastal range of mountains roughly eighteen miles due north of San Luis Obispo, California.

in that neighborhood and my inability to take an escort determined me to proceed to the valley of the San Joaquin by Pacheco's Pass.[67]

I accordingly left this place on the 5th of July, accompanied by Capt. Halleck, Secretary of State, Major Canby, Capt. Westcott, and Lieut Derby. Passing by the mission of San Juan Bautista and through Pacheco's Pass I crossed the San Joaquin river at the lower mouth of the Merced, the Tuolumne near its mouth, and the Stanislaus river at Taylor's ferry about thirty miles in a south easterly direction from the town of Stockton.[68]

Pacheco's pass was found to be of easy practicability for mules, but exceedingly difficult for wagons. As a route for the transmission of supplies from the Coast into the interior considerable labor must be expended upon it before it can be made a feasible one. The pass of San Juan a few miles to the South is less difficult for wagons but presents the additional disadvantage of being without water during the season of the year best suited to land transportation.[69] The country in the immediate vicinity of my route was found to be exceedingly barren and singularly destitute of resources. Except a narrow strip on the borders of the streams it was without timber and grass and can never in my estimation be brought into requisition for agricultural purposes. The timber on the banks of the San Joaquin, Tuolumne and Stanislaus rivers is composed almost entirely of the holly leafed oak a species of white oak, willow, and sycamore. The timber is low, dwarfish, apparently hard to work and unfit for building purposes.

Major Miller was encamped with two companies of the 2nd Infantry at Taylor's ferry in an apparently healthy situation, with fine water and

---

[67]Pacheco's Pass, described by Derby, began roughly twenty miles northeast of Monterey and led into the lower San Joaquin River valley.

[68]The route was well known by this time and was considered the quickest way to move from the coastal region of Santa Clara into the mining district. Taylor's ferry, here identified, was on the main road, located just six miles above the mouth of the Stanislaus River. See George Henry Tinkham, *History of Stanislaus County California: With Biographical Sketches of the Leading Men and Women of the County Who Have Been Identified with Its Growth, etc.* (Los Angeles: Historic Record, 1921), 77.

[69]There were two routes into the Central Valley from San Juan Bautista: Pacheco's Pass, which the party took, and one designated Santa Anna Pass, which Riley called San Juan, that led almost directly south and then east into the Central Valley. Santa Anna Pass wandered through the mountains for many miles before crossing the mountains, thus the reference to the lack of water.

an abundance of fire wood. This command was established in its present position for the purpose of preventing conflict between the Indians and the whites and to be at hand should a collision (as was then much apprehended) occur between the Americans and the foreigners, who had congregated in great numbers upon the water of the Stanislaus, Tuolumne and Merced rivers. The necessity of keeping troops in this position to meet Indian difficulties will probably not exist much longer. The rapidly increasingly mining population is gradually forcing the Indians to the South and another season will find them so surrounded by whites that the presence of troops unless it be to protect the Indians, will no longer be required. The danger of the occurrence of the last mentioned difficulty has also in a great measure passed away.

I was disappointed in finding that company "E" 1st Dragoons had not joined Major Miller's command and I was in consequence reluctantly obliged to defer my proposed examination of the country in the neighborhood of the Lake Buena Vista, until able at some future time to obtain a mounted escort.[70]

From Major Miller's command I proceeded through the mineral region on the head waters of the Tuolumne and Stanislaus Rivers to the Sacramento, crossing on my route the Calaveras, Mokelumne, Seco & Cosumnes rivers, for the purpose of examining the valley of the Sacramento river for the distance of eighty or one hundred miles above the confluence with the American Fork.[71]

The selection of a military post in that neighborhood had been ordered by the Commander of this Department in January last but the Survey was suspended by the transfer of the Topographical Engineers in this Department to Division Head Quarters.

Upon my arrival at Major Kingsbury's camp I found that the Major General Commanding the Division had already selected a position for this command on Bear Creek a tributary of Feather River and distant about thirty miles from Sutter's Fort.

The country between the Stanislaus and Sacramento is composed

[70]Derby undertook this survey in 1850, the journal of which will be presented in chapter 3.

[71]The party never went up the Sacramento, so Lieutenant Derby was ordered to undertake that survey in the fall 1849, which resulted in the development of his second journal.

of exceedingly hilly ground and dry, and plains unfit for cultivation except in particular localities on the borders of the streams constituting an infinitely small portion of the surface of the country. The timber is generally larger but of the same kinds as on the western side of the San Joaquin and the same destitution of resources obtains as on that side of the valley. As you ascend the lower slopc of the Sierra Nevada, pine of an indifferent quality becomes common but exists in situations where it is difficult of access and not in sufficient quantity to meet the demands created by the rapidly increasing population of this country. It cannot therefore be depended upon in the erection of quarters in the interior, and as before stated, the oak of the country is unfit for building purposes. For the erection of the walls of the building, adobe may be used and in the fabrication of these, Indian labor may economically be called into requisition. That portion of the valley of the Sacramento visited by me is far better suited for agricultural purposes than the valley of the San Joaquin and I learn from the reports of others that this advantage becomes more apparent as you proceed to the north.

The position occupied by Major Kingsbury was selected for temporary occupation under instructions from Department Head Quarters for the Purpose of putting an end to outrages that were being committed by the whites upon the Indians of that neighborhood. In its new position [Camp Far West on Bear River] this command may readily be supplied by water transportation.[72]

The country lying between the Stanislaus and King's river, or Lake Fork, presents the same general features, barren hills, and arid plains, with narrow strips of arable land on the borders of the streams running from the Sierra Nevada. In the neighborhood of the lake [Tache] the banks of these streams are densely populated with Indians divided into many independent tribes, or Rancherias, speaking different languages and exceedingly difficult to control.[73] Many of these Rancherias profess

[72]These "outrages" were rather minor, and conflict had subsided by the next year. See the reports from "Containment Far West" in Captain Hannibal Day to Assistant Adjutant General, May 16, 28, July 1, and November 1, 1850, NARG 393, LR, DP.

[73]Good descriptions of these Maidu and Miwok Indians appear in the Derby journal transcribed in chapter 3 as well as in the several reports of Captain E. D. Keyes. See Keyes to Assistant Adjutant General, March 19, April 11, and June 11, 1851, NARG 393, LR, DP.

to be friendly to the Americans and will probably continue so unless induced to become hostile by aggressions committed upon them. They subsist chiefly by hunting and fishing, or upon the seeds of several grasses, or upon acorns, and upon the appearance of an enemy abandon their villages and take refuge in the mountains. No disposition that can be made of troops will give security unless they be stationed in some controlling position in their midst.

Although it will be difficult to supply a garrison in that vicinity unless the San Joaquin river should be found navigable to a much greater distance than is now believed to be it will be more economical in the end than any other course that can be adopted.

The three routes by which these supplies may be sent are, from Benicia by the river and valley of the San Joaquin, from Monterey by Pacheco's Pass, or the pass of the San Miguel. The best of these routes can only be selected after a careful examination of all. [Derby undertakes this task in 1850.]

Very Respectfully, Colonel
Your Obed. Servt.
B. Riley
Bvt. Brig. Gen
Commanding

Lieut. Col. W. G. Freeman
Asst. Adjt. Genl. U.S.A.

Riley's second report contained even more detailed information regarding the Indians of the mining district. In 1849, the army's main concern was the possible hostility of the Indians living along the rivers where mining was taking place. During the following year, the army would soon become more concerned about the hostility of the miners toward the Indians.

Letter to General Persifor F. Smith

Head Quarters 10 Mil Dept
Monterey Cal. Oct. 15, 1849

General [Persifor F. Smith],

I have heretofore reported to General Head Quarters the difficulties that are anticipated in the management and control of the Indians in this department. As these difficulties are of so important a character both to the welfare of the Indians and the quiet of the frontier that the interference of Congress may be deemed necessary, I respectfully invite the attention of the Secretary of War to the following consideration.

The Indians of California are scattered throughout the whole extent of the country from north to south, but in the greatest numbers along the western slopes of the Sierra Nevada, in small bands or Rancherias without any general organization and generally without acknowledging any authority superior to that of Captains or Chiefs of their Rancherias. They are divided into three classes, the Christianized Indians or Neophytes of the missions many of whom are domesticated in the families or upon the ranchos of the inhabitants of California and are properly subject to the local laws of the country. From these no trouble is anticipated. The friendly or tame Indians (Mansitos) [here Riley uses the diminutive form of *mansos,* which means "tame"] live in small communities upon the banks of the Sacramento and San Joaquin and their tributaries living upon game and fruits. These are of a very degraded class but generally harmless and inoffensive and living in constant and friendly intercourse with the whites in their neighborhoods. The wild Indians of the Sierra, more degraded than either of the other classes, living in the slopes of the Sierra, and subsisting upon game, acorns, roots and upon the products of their thieving incursions. Among these are many renegade *Christian* Indians, who by their superior tact and intelligence have risen to a controlling influence among these Indians and are the leaders in the incursions for stealing horses and in all acts of hostility against the whites.[74] All of these Indians are greatly addicted

[74]Spaniards generally called these former mission Indians *Ladinos,* which means literally sagacious and cunning, for they often spoke Spanish and knew the location of mission livestock,

to intemperance and when in contact with, rapidly acquire all of the vices and none of the virtues of the Whites.

So many different dialects are spoken among the Indians of California that the inhabitants of Rancherias separated by but a few leagues are unable to understand each other and among the Neophytes of the same mission there are four or even a greater number of distinct languages frequently spoken.[75]

From the greater number of distinct communities the diversity of interest and confusion of languages it will manifestly be a matter of extreme difficulty to make these Indians understand the nature of their relations to our country and their white neighbors, and consequently to prevent those collisions and broils that will inevitably result from intimate contact with our people, and from the encroachment of the unprincipled particularly as there are no Laws now in existence that will fully meet the peculiar situation of these people.

The Laws of the United States regulating intercourse with Indian tribes generally contemplated an Indian Territory over which Congress exercises an exclusive sovereignty, but there is in California no recognized Indian territory, and the Laws of California will unquestionably, so soon as it becomes a state, be extended over all the territory west of the *Sierra Nevada.*[76] The wise and humane provisions of our Indian laws will thus be rendered entirely nugatory or the officers of the General Government charged with the execution of those Laws be brought into direct and constant collision with the state authorities.

---

which they raided. (Of course, Ladino is also the language spoken and written by Jews of Spanish origin.) For a history of the efforts of the Bureau of Indian Affairs to deal with these groups, see George H. Phillips, *Indians and Indian Agents: The Origins of the Reservation System in California, 1849–1852* (Norman: University of Oklahoma Press, 1997).

[75]Most of the Indians living in the Central Valley, and in the Sierra foothills between the Cosumnes River in the north and the Fresno River in the south, were generally referred to as Miwoks, a Native term that means literally "the people." They then broke down linguistically into roughly four groups, with many different dialects. See Kroeber, *Handbook of the Indians of California*, 442–73.

[76]The problem that Riley was addressing stems from the Intercourse Act of 1834, which defined "Indian Country" as that land between the Mississippi River and the Rocky Mountains, excluding Missouri and Arkansas. Thus, California was not a part of "Indian Country," and not subject to the law. See Francis Paul Prucha, *American Indian Policy in the Formative Years: The Indian Trade and Intercourse Acts, 1790–1834* (Lincoln: University of Nebraska Press, 1962).

Unless immediate measures be taken by Congress to avert this evil I apprehend a recurrence of those misunderstandings which a few years since occasioned so much trouble in some of our southern states.[77] This evil too will exist in California in its most exaggerated form for there is already here a border population collected from all parts of the World who eagerly seize upon the slightest pretence [*sic*] to give expression to their hostility to the Indian race.

It is suggested that the title of the Indians to the lands of which they may be in actual occupancy be immediately recognized by Congress and that measures be taken to remove them from the immediate vicinity of the white settlements and to establish them in some suitable location to the east of the *Sierra Nevada,* or if no suitable territory can be found, and it is extremely doubtful whether it can, that the exclusive jurisdiction of the United States be reserved over sufficient territory west of the Sierra to admit of the concentration of all the Indians not already domesticated among the inhabitants of California: as suitable districts for reservations the country immediately east of the Lake Buena Vista (Tulares of Frémont) [reference here made to Frémont's map] extending to the crest of the Sierra Nevada and that around the laguna [Clear Lake], north east of Sonoma (& including Russian River) are suggested. These districts are more densely populated by Indians than others in California and probably will not be required by the increasing population of the country until sufficient time has elapsed to make some permanent arrangements for these Indians.[78]

The dictates of humanity require that prompt and efficient measures should be taken to prevent a further demoralization of the Indians of this country, and this cannot be done effectually unless they be separated from the white population in their neighborhood and the

---

[77]Riley is here referring to the great debate in Congress over Indian Removal. As a consolation to mainly northern congressmen who opposed the removal of southern Indians, the Intercourse Acts were passed to supposedly give Indians west of the Mississippi River a clear title to their lands.

[78]The Native populations of these areas were not well known until the various surveys were completed by the army, including the four undertaken by Derby and in particular the survey of the Russian, Eel, Klamath, and Trinity river valleys in 1851 by Major H. W. Wessells, whose journal is reproduced here in chapter 4. The populations were much smaller than local Anglo inhabitants believed.

traffic in spirituous liquors be prevented. Provisions should also be made for their support until they are so far civilized as to be able to support themselves wholly, or in great part, by the cultivation of the soils, for if surrounded by a population of whites as they unquestionably soon will be if the extraordinary immigration now in progress continues even for a few months, the land in the immediate vicinity of their Rancherias will not furnish them with subsistence and they will perish from starvation or be forced by want to commit depredations upon the property of their neighbors, and the commission of these depredations will be the pretence [*sic*] for the commencement of a War of Extermination.

Very Respectfully General
Your Obdt. Servt.
B. Riley
Bvt. Brig. Genl, U.SA.
Comd. the Dept.

Edward O. C. Ord's Journals of the Basin of Los Angeles October–November 1849

While Riley and Derby were surveying the mining district to the east of the Central Valley, Captain Edward O. C. Ord spent the fall examining the land around Los Angeles, to the east of it, including the San Bernardino Mountains, and to the south. By December, he had created a map of the Los Angeles basin, and his two journals, reproduced below, gave careful detail to the various geological features of the region, the passes into the interior, which Indians had used for many decades to attack the California ranches along the coast, and the population. His charge was to locate possible sites for military posts, in order to end the incursions of Indians, who often came into the ranching areas along the coast from the Tule River region.[79]

[79]The first Ord journal is found in Captain Ord to General Riley, October 31, 1849, NARG 393, LR, DP. This journal was also published in Senate Executive Document 47, 31st Congress, 1st session.

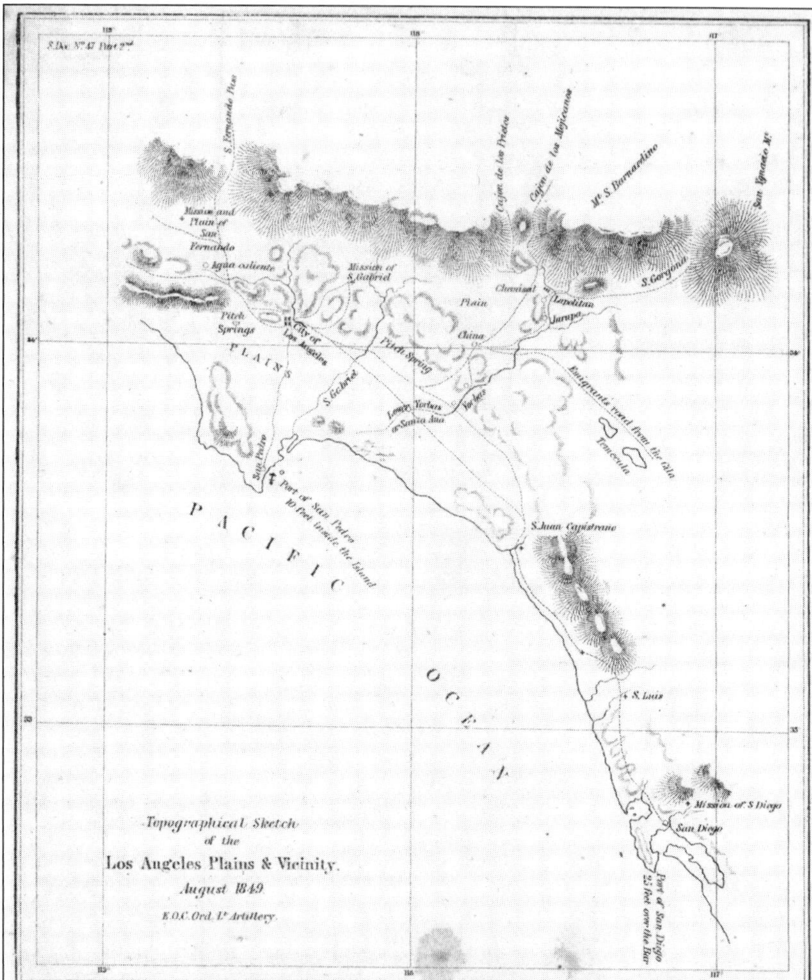

"Topographical Sketch of the Los Angeles Plains and Vicinity, August 1849."
From Edward O. C. Ord's journals of the army survey of the Los Angeles basin, October–November 1849.
Courtesy of National Archives and Records Administration.

Letter to Major E. R. S. Canby

Monterey Cal. Oct 31, 1849

Sir:

In obedience to orders I left this place in the steamer [not named] of the first inst. Landed at San Diego on the 3rd and next afternoon (as soon as horses could be got) proceeded over some seventy five miles of clay hills, by the shortest road to the plains of Los Angeles. These plains have an average width of about thirty miles, length about seventy-five—watered by three short rivers running from the mountains to the sea in winter, and sinking in the plains in summer. Here are the principal cattle ranchos of the Lower Country into which the Indians enter to steal. 1st from the Pa-utah country on the east by the Cajon—2nd from the Tulare or San Joaquin country by the San Fernando pass, & 3d from the Cowilla country on the Southeast by the San Gorgona and Warners passes.[80] As the Pa-utahs are the greatest thieves and most cattle are driven out by the "*Cajon de los Mexicanos*," I turned up the Santa Ana and proceeded at once to the Rancho of Col Williams, examined its houses and grounds and next day crossed the plain between this rancho and the Cajon to "Jurupa" on the Santa Ana, 21 miles—thence followed up the river to Lapolitana via the Mexican Pueblo, 9 miles—thence up Cajon Creek to the pass 10 miles, thence two or three miles in the pass.[81] At the mouth or entrance of the pass

[80] *Pautah* was a term that appeared on the Frémont map. It generally extended throughout the Sierra Nevada, from the American River southward to Walker's Pass. The term also often referred to the Indians of that region. Pautah County was briefly a California County on the eastern slope of the mountain, but the land was later given to Nevada. San Fernando Pass was located northeast of Los Angeles, separating the Santa Susana Mountains from the San Gabriel Mountains. Warner's Pass, named after John J. Warner, was on the southern end of Pautah. Warner, who received a large Mexican land grant, used the local Cupeño Indians as laborers on his ranch, keeping them in a condition of near serfdom. They rebelled in 1851, burning Warner's ranch. Warner's Hot Springs is also named after him. See Richard W. Amero, *Eviction at Cupa*, at http://balboaparkhistory.net; Edward D. Castillo, "The Impact of Euro-American Exploration and Settlement," in Heizer, ed., *Handbook of North American Indians*, 8: 110.

[81] Ord is just north of present-day Riverside, marching along the Santa Anna River, moving into the Jurupa Pass. Isaac Williams came to California in 1832, married the daughter of Antonio Maria Lugo, and took over Lugo rancho, which was located near present-day Chino, California. He was often questioned by the army about his knowledge of California geography. See Robert Cowan, *Ranchos of California: A List of Spanish Concessions, 1775–1822, and Mexican Grants, 1822–1846* (Fresno, Calif.: Library Guild, 1956).

and for a mile or two on either side, the land though flat is stony and barren, and for four or five miles out in the plain without grass or covered with "Chemisal" or low bushes. There is water in the pass but not grass enough to support a picket of cavalry and even should one be posted here it would not prevent the entrance of the Indians, who could easily avoid such picket by a detour of about a league to the right by another though not very good pass called "El Cajon de los negros"; after examining the pass I attempted to cross the plain in a direct line thro' the chemisal to Williams' but night came on, my guide got into some hills on the left, got lost and I had to lead him to Bandini's rancho on the Santa Ana—next morning returned to Williams', found the buildings ample for one company, the position facing the cajon, 18 miles off in intervening hills and no time to put up quarters on the Indian side of the pass. I contracted for quarters there (see contract), after which I rode some eighteen miles to reported coal beds; found it Bitumen, hardened.[82] Next day I returned to Bernardo Yorba's the largest and most reliable farmer in the neighborhood, contracted for three hundred fanegas of corn (all he could spare), I found no one else in the vicinity willing to make large contracts & I deemed it best that small purchases should be left to the quarter-master.[83]

From Yorba's I pushed on to the Pueblo, making inquiries as to the disposition of the people to defend their property in case arms and ammunition should be issued them. They are willing and if properly organized would be able to do so against all Indian incursions, but there is no unity of action amongst them.

At the pueblo I spoke to the judge, the alcalde, Messrs Wilson, Temple, Pio Pico, Manuel Riquena and other of the principal inhabitants urging the formation of a volunteer company, or some effort at organization that they might avail themselves of arms etc., to be issued

---

[82]Juan Bandini's land grant, dated 1839, included four thousand acres in the vicinity of Chico, California. Ord covers this ground again in November, in the journal that follows, with more detail.

[83]There were four Yorba brothers, all with ranches in the Los Angeles basin. Bernardo's ranch was in Santa Anna Cañon, roughly eight miles east of Placentia, California, some fifteen to twenty miles southwest of where Ord proposed quartering cavalry. See Terry E. Stephenson, *Don Bernardo Yorba* (Los Angeles: Glen Dawson, 1963), 2–3.

them.[84] They all acknowledged the propriety of such a measure, said they would consider it and would no doubt form some company. I think a supply of arms, placed in charge of some responsible person to be issued when required, or upon the organization of their community, an advisable measure under present circumstances.

As I was anxious to get back in time to allow such measures to be taken as might be thought proper, before the rainy season, I returned to San Diego, waited there for the Steamer until the 28th when, being impatient, I hired horses and was about starting up by land, when the steamer hove in sight; I got on board & arrived here this morning. Allow me to state that though William's rancho appears to be the best point in the vicinity of the Cajon to quarter troops before the rainy season sets in, I do not deem it the proper place for a permanent garrison to protect that district from the Indian horse thieves. 'Tis reported to be but two days with wagons through this pass to the Pa-Utah country on the Mojave, the valley of which is large, well watered with good timber for building but, 4 or 5 miles from the water course and abundance of fine grass; 'tis on the most direct route for New Mexico & the Salt Lake in the plain, upon which open three important & easy passes—that to the north into the San Joaquin plains—that to the east leading to the Great Basin & Indian country of the Apaches, Utahs & Navahoes—and the cajon to the Pacific. The upper end of the San Joaquin district where the San Fernando and Tejon passes enter would require other defences [*sic*]—the San Gorgona and Warner's passes appear to require no permanent establishment—the horrors of the Gila route will prevent its use after the present year and the Indians in that quarter are not *yet* hostile.

[84] Ord mentions some of the most important ranchers in the area surrounding Los Angeles. Jonathan Temple came to California in 1828, married a Mexican woman, and developed Rancho Los Cerritos. He was the first Alcalde of Los Angeles after the Americans captured it. Pío Pico was the last Mexican governor of California, who tried to rally Mexican forces after the American invasion. He fled to Mexico, then returned to his ranch in July 1848. He was briefly arrested, released, and thereafter served on the Los Angeles City Council during the 1850s. Manuel Riquena also served as alcalde of Los Angeles and later on the city council. See Carlos Manuel Salomon, *Pío Pico: The Last Governor of Mexican California* (Norman: University of Oklahoma Press, 2010), and Governor Richard Mason to Colonel R. Jones, August 23, 1848, House Executive Document 17, 31st Congress, 1st session.

I further report that many emigrants arriving at San Diego are destitute—neither bread, flour, nor sugar can be purchased there. I requested the commanding officer to allow sales in small quantities to them until your orders in the case could be received. The sketch to the above report will be handed in as soon as completed.[85]

Very Respectfully Etc
(signed) E. O. C. Ord

To
Maj E. R. S. Canby
Asst Adjt Genl, USA

ORD'S SECOND JOURNAL OF HIS RECONNAISSANCE OF THE LOS Angeles basin is included below. It is unclear as to why the second trip was made, other than to answer certain questions that came up regarding Cajon Pass. Ord also spends considerable time in and about San Diego, giving a detailed sketch of the town and the mission that had existed farther inland.[86]

Letter to Major E. R. S. Canby

Monterey
Nov 6/ 49

Sir:

In compliance with part of my directions received from Departmental Hd Qrs. in connection with my reconnaissance of the Cajon Pass, I have the honor to report—

That I landed at San Diego; here there is a good frigate Harbor, the quietest on this coast—no good survey has yet been made of it except by the English—one such by us is very much wanted as vessels from

[85]The "sketch" is undoubtedly a reference to the 1849 Ord map, which is included here.

[86]The original journal is found at Ord to Canby, November 6, 1849, NARG 393, LR, DP. Reprint found in Senate Executive Document 47, 31st Congress, 1st session.

China and the Islands make this part of the coast first and have to beat up to San Francisco against the prevailing northers—the Steamers have established their coal depot at this place in preference to San Francisco. The entrance, about two hundred yards wide is of easy defence [*sic*]. The town containing about twenty badly built adobe houses and two pretty good ones, is five miles from the anchorage; a road leads around the Bay shore, bordered by naked hills—small boats cannot get within half a mile of the town owing to shallow water and march mud. The town was built here because it's the only point on the Bay where a good supply of water can be had from the little river which here in summer empties its waters into the Bay—in autumn these waters dry, or sink into the Sandy bed. I followed this bed of sand through its valley amongst low cactus-covered clay hills, seven miles to the mission. There I found the old walls tumbling in and everything going to ruins. The vineyard and olive trees still produce a little fruit. The library should be placed in charge of some one interested in its preservation. From the mission along the river bed 'tis about five miles more to a gap in the hills where the valley heads, near this gap are a few sycamore trees & willow bushes, the only timber in the valley. This valley though much smaller than that of the Carmel [south of Monterey], has a good soil and if cultivated would maintain perhaps three or four hundred inhabitants. There is still in good preservation an aqueduct leading from the running water in the hills to the mission, which with a little care would supply water for irrigation to the whole valley. The Rocks at the gap above the mission appear to be of trap recently upheaved [*sic*]; between this point and the beach, the gravel and clay hills are composed of a sedimentary deposit—coal was reported to exist near San Diego, but the late Captain Warner upon examination found it to consist of a small deposit of Lignite.[87]

The Bay shore south of San Diego is bordered by the same low gravel & clay hills already described—no timber—but little grass—plenty of cacti occupied by hares, rabbits & quails—

From San Diego I took the road N. West up the coast, toward Los

[87]Brevet Captain William Warner, of the topographical engineers, was exploring the region east of Goose Lake in northern California in September 1849 when Indians ambushed and killed him. The Indians involved were never identified.

Angeles, which road passed round the edge of False Bay, a shallow arm of the sea about three or four miles N. of San Diego—no hills between them, though a bluff separates the two bays near the sea. From the False Bay the road for thirty miles leads amongst the low table hills of the coast, similar formation to those above described and equally as barren, though at intervals of from four to ten miles it crosses little valleys, which when watered contain an adobe hut or two, some small patches of corn, beans, & pumpkins, and a few cattle. The valley of San Dieguito, 20 miles from San Diego, if all cultivated would produce several thousand *fanegas* [about a bushel and a half] of maize, and other things in proportion, from the half dozen patches I saw there, the Indians were gathering from twenty to forty *fanegas* of maize to the acre—this without irrigation. About thirty-six or forty miles from San Diego in a similar valley, near the sea side, the road passes the now deserted mission of San Luis Rey.[88] It was well built, and its handsome church, extensive arcades, orchard and vineyard all look fresh, though the people are all dead and gone away. I found that the birds and the sunshine had entered by a hole in the roof and one little gorion [suggesting a learned "sage" or monk] was perched over the altar, singing away as if the place belonged to him. Some one interested in their preservation should be sent there to protect the woodwork & gardens from the ravages of travelers. Three or four miles to the north of San Luis is another and better watered valley, occupied by a few Indians and Don Pio Pico's rancho. This gentleman has a pretty good vineyard though he cultivated but little—the valley would maintain some thousand souls—near this valley where the road crosses a spur of rocky coast hills, I observed what I took to be marl or rotten lime stone.[89] Five miles from the valley the road again strikes the beach near the little mission

---

[88]Mission San Luis Rey de Francia was built at present-day Oceanside, California, in 1798. The mission church, completed in 1811, has a unique architecture, a combination of Spanish Renaissance, Moorish-Mudéjar, and Spanish colonial forms. A neophyte, Pablo Tac, composed a history of his life at the mission, written in Rome about 1835. See Lisbeth Haas, ed., *Pablo Tac, Indigenous Scholar: Writing on Luiseño Language and Colonial History, c. 1840* (Berkeley: University of California Press, 2011), as well as the older Minna and Gordon Hewes, eds., *Indian Life and Customs at Mission San Luis Rey* (1958; reprint, Whitefish, Mont.: Literary Licensing, 2012).

[89]Pío Pico had several ranches located between San Diego and Los Angeles; the reference here is to Rancho Jamul. The Indians were former members of the mission, just four miles west of the ranch. See Salomon, *Pío Pico,* 54. Marl is generally identified as lime-rich mud containing clay and silt.

of Las Flores, as usual going to ruins.[90] Thence it follows the beach for twenty or twenty five miles to the valley of San Juan Capistrano—keeping aforesaid coast range close in the right. These rocky hills appear to be of the same formation as the volcanic ridges in the Stanislaus gold region. I observed a prevalence of talcose slate [largely talc] interspersed with quartz veins and boulders, supersoil [*sic*] of red-clay and igneous gravel, all of which are characteristics of the gold soil. Gold is said to have been found here.

In the pretty little valley of San Juan are the remains of the mission and church of that name.[91] The latter was built with an effort at ornament as the handsome cut-stone cornices, pilasters, & medallions still show. During the great earthquake of 1814—one half of this building fell, and crushed to death a number of persons in its fall. There are still eighteen or twenty four families living in the valley. They raise an abundance of corn, pulse [peas, beans, or lentils] etc. and some very good peaches.

From San Juan the road turns inland and leads north and for six or eight miles through low prairie hills to the South end of [the] Los Angeles plain. These plains extend from San Juan on the S.E. to *Canenga* [*Cahuenga*, near present-day Burbank] on the N.W. a distance of about seventy miles; on the left the coast & harbor of San Pedro (no survey of them has yet been made except by the English) on the right a range of low hills, and farther inland the face of the Sierra Madre and San Bernardino, from which flow into the plains three small rivers, the Santa Anna, the San Gabriel and Los Angeles.[92] Their waters last with but little intermission, tho' in the fall they sink in the sand, to rise again lower down the plain in sundry Springs and wet places. In the Spring these plains are covered with grass, geese, flowers and cattle. The grass dries in the summer the geese fly away, the flowers die and the cattle remain to fatten on the hay. Formerly they used to kill them in the fall

[90]Mission Las Flores Estancia was a "station" about halfway between San Luis and San Juan Capistrano.

[91]Mission San Juan Capistrano was constructed in 1776, the chapel being completed six years later. At its zenith, the Native population reached a thousand. It also has the distinction of being the oldest building still standing in California today, and the mission's vineyard was the first to produce wine. See Zephyrin Engelhardt, *San Juan Capistrano Mission* (San Francisco: Standard Printing, 1922).

[92]Rancho Cahuenga was granted to José Miguel Triunfo in 1843. Triunfo was perhaps the only American Indian to receive a grant—it was only about four hundred acres—as a reward for his work at San Fernando Mission. It was located in present-day Burbank, California.

for their hides and tallow, but now there are no more hide-droguers [derivative of drogue, a contrivance attached to the end of a harpoon line] and the cattle have to perform long journeys north to carry beef to the mines. I skirted the plains near the hills on the right to where the Santa Ana came in (passing one rancho or cattle farm). Shortly after entering the plain the waters of the Santa Ana sink in its sandy bed. Here the road continues on to Los Angeles, but I turned up the sandy bed of the Stream and in half an hour's ride, the valley and its grove of trees opened before us, at the mouth of which valley on the ranch of Theodosio Yorba, are three or four snug houses, a good vineyard and orchard and about ten or twelve acres under cultivation (he might have ten or twelve hundred) from which he expected three or four hundred *fanegas* of corn.[93] Here I found the river a very respectable stream, crossed it where it had a rapid current—width of sixty to a hundred yards, and a depth of from 6 to 20 inches of sweet clear water. 'Tis the largest river south of San Francisco bay in California, and though I had been nearly three years in the country I did not think there was such a pretty valley in it as I now entered.[94] 'Tis from one half to two miles wide, low verdant banks covered with groves of cotton-wood, willow, sycamore. An hour's gallop carried me to the house of Don Bernardo Yorba, large clean and comfortable.[95] He has a large vineyard and thrice as much land under cultivation as his brother Theodosio, couldn't say how much corn he'd father as he did not expect to get it in before November. 'Twas then 9th of October. The people here frequently plant maize in July and August, when they are eating roasting ears from the first crop. In this and the neighboring valleys, owing to the abundance of water and to these waters being nearly even with the

[93]Theodosio (or Teodocio) Yorba was the youngest of the four brothers, born in 1809. He was a close friend of Pío Pico, the last Mexican governor of California, and received several large land grants from him. He fell into financial trouble in the late 1850s and was forced to sell his Rancho Lomas de Santiago, along the Santa Anna River, to William Wolfskill, who planted thirty-two thousand vines and was producing fifty thousand gallons of wine a year. Wolfskill, a former trader, likely became the richest man in California and is often thought of as the father of the wine industry. See Stephenson, *Don Bernardo Yorba,* 15–17; Engstrand, *William Wolfskill.*

[94]The Santa Anna River.

[95]Farther up the Santa Anna River, Yorba owned two ranches, El Rincón and La Sierra. Ord visited Yorba's ranches again in 1856, reporting that a small community of Indians still survived on them. See Harlow, ed., *City of Angels,* 21. The Huntington Library has a copy of Ord's 1856 journal, the original of which is also in NARG 393, LR, DP.

river bottoms through which they flow, irrigation is but trifling labor, and the farmer need not fear losing his crop by too much rain.

From Bernardo Yorba's I took a short cut across the hills to the Chino rancho, eight miles off. From the top of these hills I had a fine view to the rear of the plains—the coast and the islands abreast; in front at my feet lay Williams' buildings & rancho, at the edge of a bright green little plain, eighteen or twenty miles across this plain rose the lofty Sierra.[96] It looked a half an hour off, & took me all day to ride it. The gap called the *Cajon* pass, or the *Cajon de los Mexicanos* was distinctly visible. Course from the hill-top N.E. to the left of this gap about a league was another also leading into the Puites [Paiute?] country, called the "*Cajon de los Negros.*"[97] To the right of the main or Mexican *Cajon,* rose the San Bernardino, the only mountain west of the Sierra Nevada on which the Snow lays till August, and which can be seen from Los Angeles, sixty miles off. Still to the right of San Bernardino my guide pointed out the San Gorjon [San Gorgonio] a pass leading into the Corvilla [Cahuilla Indians] country & on towards Red River.[98] The range to the West of the *Cajon de los Prietos or Negros,* is continuous like a great wall for about fifty or sixty miles to the pass of *San Fernando,* or the "*Cajon de las Uvas,*" this leads to the head of the Tulare or San Joaquin plains—to the East of the *Cajon de los Negros* the range is broken into isolated peaks with intervening valleys.[99] To the S. east was seen the emigrants road coming in from the Gila and the desert—directly through the *cajon* [San Fernando Pass] led n. easterly the way to the Great Basin and Salt Lake. After taking the bearings of all these I descended to Col. Williams' rancho and a good breakfast. The Col. has a large and pretty place, which being on the lower or damper side of the valley preserves its greenness late in the fall. He has a good orchard and vineyard and I have forgotten how many thousand *fanegas*

[96]The "Chino" Ranch belonged to Isaac Williams, who came to California in 1832 with Ewing Young. The ranch was located just south of present-day Pomona and Ontario. Ord visited it again in 1856. See Ord, *The City of the Angels,* 13.

[97]The printed Senate version of the report gives "Pintes Country," but the original in Ord's hand looks like "Puites."

[98]Ord gives descriptions of these passes in his previous report of October 31, 1849.

[99]Cajon de los Prietos translates simply as Black, or Narrow, Canyon. Cajon de las Uvas translates literally as Canyon of the Grapes, but it was commonly called San Fernando Pass. Today, it follows Interstate 5 north out of Los Angeles.

of wheat he said he could raise without irrigation. This place being the crossing point of two important routes may be the site of some future town. I rode across the south side of the valley striking the river again at Bandini's Rancho—thence I skirted along the river bank two or three miles, where I struck another spur of hills which abutted on the river; crossing these I descended into the river bottom at a place called Jurupa, a flourishing Rancho about twenty miles from Williams'.[100] Thence following the river, still bordered with groves of cotton-wood, sycamore, walnut and willow, I passed a settlement of New Mexicans about a league above Jurupa & a couple more leagues brought me to where the river forks—the left hand fork called *Cajon* creek—on this is a flourishing settlement of Couilla [Cahuilla] Indians—about forty lodges. They were getting in their corn and I was pleased to see that their industry had been rewarded for they were housing about six hundred *fanegas* of fine corn & plenty of pumpkins.[101]

I followed *cajon* creek N. westward about two leagues, until arriving opposite the *cajon* where the land became very stony & sterile, covered with the debris washed from the mountain gorges. Here I turned into the *cajon* and made the examination already reported. On my return my guide in attempting to cross the plain got benighted and took to the hills on the S. East of the valley, the spur of which I had crossed near Jurupa. I noticed in galloping through three hills large boulders of well-worn granite rolled from their beds, and well-worn corners of the same projecting from the earth evidently of an antediluvian birth—also boulders and veins of hard limestone. Next morning I got back to Williams' and spent the day in scouring the hills between his place and the coast, with one of his assistants, after coal which he said he had lately discovered. We couldn't find the coal but we got stuck in several pitch springs, which article abounds in this formation all

[100]Bandini, or Jurupa, Ranch belonged at one time to Juan Bandini. Portions of it had already been purchased by Benjamin D. Wilson. When Ord visited the region in 1856, he noted a goodly number of Cahuilla Indians living on the ranch. See Ord, *The City of the Angels*, 23n37 and 24.

[101]Ord is here referring to the Cahuilla Indians, who were at one time likely connected to a mission substation at San Bernardino. The term's meaning is uncertain, but it may originate from the Cupan language, meaning "master" or "boss." Nearly a dozen bands of Cahuilla people have survived today and have small reservations in California, extending from the region south of San Bernardino to Warner's Hot Springs, just northeast of San Diego. See Kroeber, *Handbook of California Indians*, 692–93; Lowell John Bean, "Cahuilla," in Heizer, ed., *Handbook of North American Indians*, 8: 575–87.

along the coast. These hills are of red sandstone and overlying clay slate dip about 70° Seaward—both strata much shattered by earthquakes,

Next morning I started for Los Angeles via Santa Ana, the road from which river led across the plain near the hills (passing three or four grazing farms) to the river San Gabriel, 25 miles. The waters of this stream run a few miles into the plain before sinking. It is similar to the Santa Ana, but with a shorter course, less water, and on the banks in the plain, better land.

On this river are the ruins of the old mission of the same name, formerly one of the richest in California. There are still cultivated some fields of corn and few vineyards, tho' the farmers' wealth here, as elsewhere in California consists in his stock which requires no labor.[102] From the San Gabriel it is about eight miles further to the town of Los Angeles, where the river of the same name comes out into the plain. For about four miles up and down this stream the river bottom is covered with vineyards, orchards, and gardens through which wind lanes leading to town. This latter is built close under the hill on the north side of the valley and consists of an old adobe church, about a hundred adobe houses scattered along a dusty plaza and three or four broad streets leading thereto.

This place, until the discovery of the Placer, by its tropical climate, abundant water and fertile plains, had attracted a population from the northern provinces of Mexico of about three thousand—amongst whom were a few retired trappers and a number of vagabond Indians, the refuse from the missions—to give an idea of this as well as other parts of the Californian population, I may as well sketch the rise and fall of the above named missions.

This whole coast a little less than a century ago was a howling wilderness covered with wild Indians and animals. Spanish missionaries came, gathered the Indians and tamed them, set them to building big churches and snug huts—made them cultivate the soil and raise great herds of cattle—their lands extending from mission to mission covered all the coast from the sea back to the mountains—and each mission had from two to six thousand proselytes—to take care of these they sent for Spanish soldiers—a corporal's guard for each mission

[102]A reference to San Gabriel Arcángel Mission, founded in 1771, located nine miles east of the center of Los Angeles on the Río Mondo, which provided abundant water for fields and the mission.

came; unluckily they brought wives and daughters—everything in the country bred well and before the Mexican Revolution was felt here, a mongrel population of rosy cheeked buxom girls, hard riding, hairy headed boys began to grow up—the fathers of these wanted some place for their cattle and families and to prevent confusion the Padres assigned them parts of the pretty vallies of San Jose to the north and Los Angeles to the South—These Pueblos called the Upper and Lower were increased from abroad and multiplied at home and when they heard of the revolution, delegates were sent to Mexico; soon the missions were secularized, i.e. the Spanish Padres turned out, Mexicans turned in; the cattle and what Indians did not run away were divided up amongst the descendants of the corporal's guard. About this time hungry blue eyed men who had wandered from the east after Beaver, came over the desert and down the mountains into these pleasant plains, where they found plenty of fat beeves and pretty girls. Unluckily again these latter preferred the new comers and the new comers became good Catholics and fathers of large families. When one of these Beaver trappers goes off and don't come back, others thinking he has found a good place, follow; so that since the arrival of the first trappers from the east, tracks have led this way until the new comers and their countrymen (come by sea) outnumbered the descendants of the corporals' guards—but, more polite than the padres who took the Indians' lands, or than said descendants, who took the Padres' lands, they sought unoccupied ground and have driven no one out whom they found in quiet possession—as nearly all the good lands around Los Angeles were occupied before they began to come in, but few blue eyed people are found there, and until these broad plains are settled by a race of men accustomed to the gratification of more wants, they will remain as they are, almost unproductive in the hands of those satisfied with the gratification of a few.[103] I think the first class of men will buy up these lands when they shall discover that attached to these plains at San Pedro is a good harbor for coasters and steamers—that the sea

[103]Ord's sentiments reflect his obvious Anglo bias. For a more objective assessment, see Ramón A. Gutiérrez and Richard Orsi, eds., *Contested Eden: California before the Gold Rush* (Berkeley: University of California Press, 1998); James J. Rawls and Richard J. Orsi, eds., *A Golden State: Mining and Economic Development in Gold Rush California* (Berkeley: University of California Press, 1998).

is quiet in summer, navigable for open boats, and that now a good market is at hand on the plain for all the vegetables produced on them at three times the prices paid in any other part of the United States.

I reckoned about ten grazing farms on the Santa Ana containing in all about 40,000 cattle and 10,000 horses about the same number on the Los Angeles and about twice that number on the San Gabriel—where are from sixty to a hundred corn fields and as many vineyards around Los Angeles producing from three to four thousand *fanegas* of corn & as many barrels of wine. The Port of San Pedro is about twenty five miles off gentle slope down the plain whole way, very good building stone in the harbor and all the hills next the plain—

While examining the country in the vicinity of the C*ajon* pass, I was only prevented going through it and examining the entrances to the Great Basin by want of time. I have however made many inquiries upon its occupants and resources, and would suggest that an examination be made by a strong force of the Indian country between the Mojave & Santa Fe as soon as possible. I am convinced that if a line of forts can be established across this country it will be the means of opening a short winter & summer route between the U.S. on the E and California on the W—the country is reported habitable, and to have an outlet near Jemes, N. Mexico for wagons. If such an outlet exists, the Gila route, now covered with carcasses and lined with graves, will be abandoned. We will hear of no more cannibalism in the Snowy Mountains, and Americans in this country will still feel and act as if they were still in the old United States.[104]

Monterey, Cal
Nov. 6, 1849

Very Respectfully,
Your Obdt Servt.
(signed) E. O. C. Ord

To Maj. E. R. S. Canby
Asst. Adjt. Genl

[104]The Pueblo of Jemez, New Mexico, is located on the road just west of Santa Fe. It had a hot spring, which made it an ideal place for rest and recuperation before heading west into California. Ord was suggesting a much more northern route, rather than the southern route through Fort Yuma on the Colorado River. The reference to cannibalism is from the Donner Party's terrible experience of being snowbound in the northern Sierra Nevada in 1846. See Ethan Rarick, *Desperate Passage: The Donner Party's Perilous Journey West* (New York: Oxford University Press, 2008).

CHAPTER 2

Derby's Journal of the Sacramento River Expedition 1849

DERBY'S SURVEY OF THE SACRAMENTO VALLEY OCCURRED OVER roughly two months during the fall of 1849. His main mission was to help locate a proper placement for Camp Far West, which was eventually laid out along Bear Creek. But the expedition then moved northward, well up Butte Creek and the Feather River, exploring the land and its occupants. Again, Derby found very few Indians, and many of them were suffering from disease. The same was true for many of the Anglo settlers who had reached the region.

The journal is preceded by a brief letter that explains the purpose of the trip and the official orders, written by Major Canby.

Orders to Lieutenant Derby

Geo H. Derby
Lt. Top. Engineers
Head Quarters, 10th Mil Dept.
Monterey Cal. September 5, 1849

Sir:

Major Kingsbury 2d Infantry was instructed from Division Head

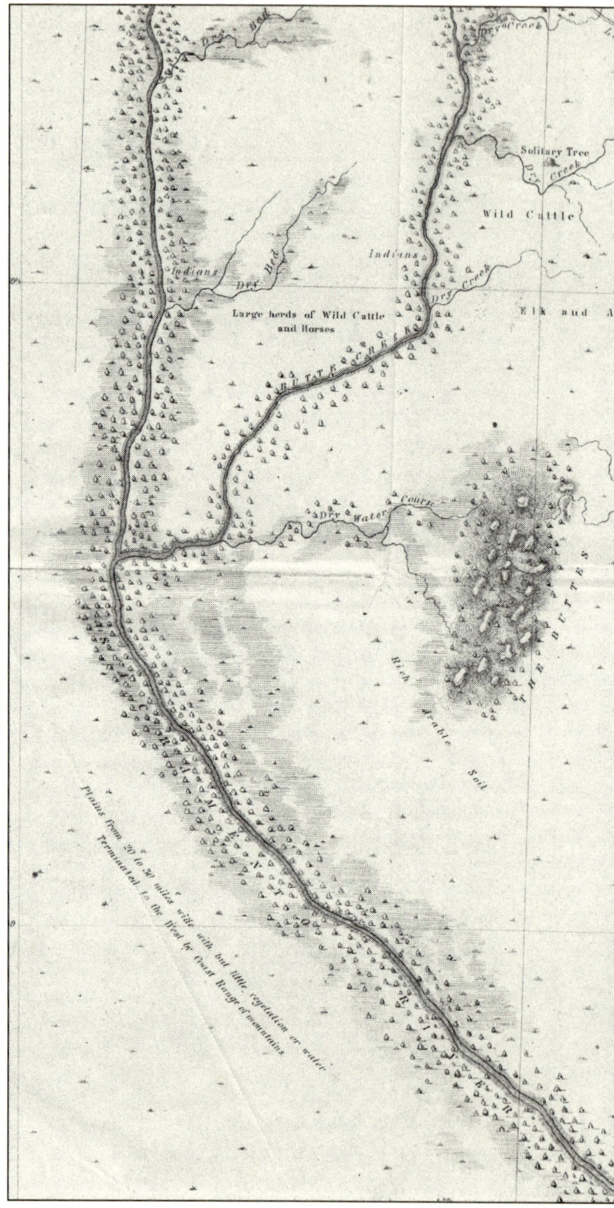

Dry Creek
Solitary Tree
Dry Creek
Wild Cattle
Indians
Indians
Dry Bed
Dry Creek
Large herds of Wild Cattle
and Horses
Butte Creek
Dry Water Course
Rich Arable Soil
Plains from 20 to 30 miles wide with but little vegetation or water
Terminated to the West by Coast Range of mountains

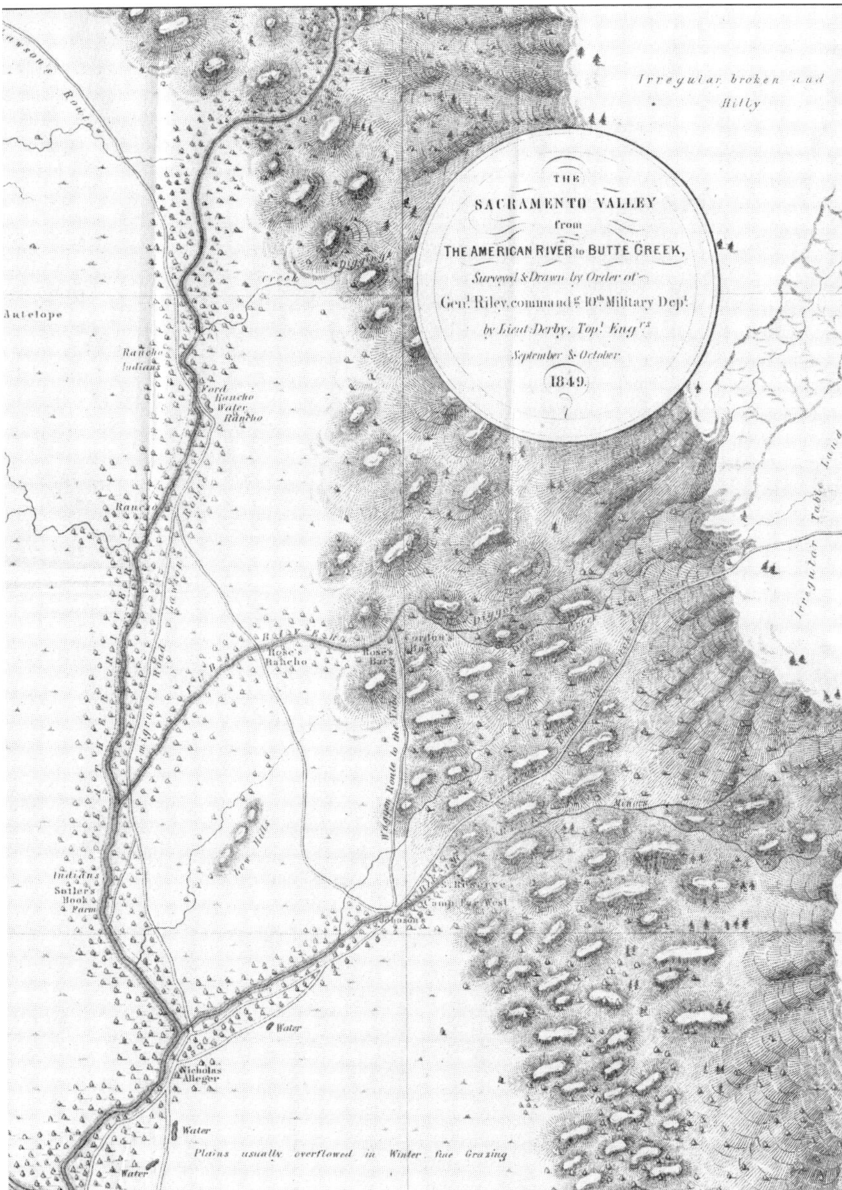

"The Sacramento Valley from the American River to Butte Creek," drawn by George Horatio Derby, September–October 1849. From Derby's journal of the Sacramento River expedition. *Courtesy of the Harold B. Lee Library, Brigham Young University.*

Quarters on the 10th of July last to establish his command at Johnson's Rancho, on Bear Creek a tributary of Feather River and to lay off a reservation of one square mile etc.[1] You will accordingly after joining Maj Kingsbury proceed to survey and mark out this reservation, and you are authorized to call upon Maj. Kingsbury for any assistance that you may require for that purpose.

Two copies of your map will be made one of which will be left with Maj. Kingsbury for the post, and the other will be forwarded to Department Head Quarters.

The commanding General has learned unofficially that Major Kingsbury has been authorized by General Smith to select any point he may deem preferable to that indicated in the instructions above referred to. In this case you are directed to render Major Kingsbury any assistance that may be in your power in enabling him to make this selection. You will at the same time make a Topographical sketch of the country that you may examine for this purpose, and collect and report for the consideration of the commanding General any information that may be useful with regard to the resources of the country, the means of communication, the number of Indians, distinguishing between the tame Indians of the Ranchos and the wild Indians of the Sierra; the comparative advantages of different positions for military posts, having particular reference to healthy locations etc. etc.

In reporting upon the resources of the country, you will designate particularly those that may be useful to a military command, such as forage, building materials, etc.

After the completion of these duties with Maj. Kingsbury's command, you will make an examination of the valley of the Sacramento, to about the latitude 39° 20' or the mouth of Butte River [Creek], collecting and reporting any information that may be useful upon the subject above referred to.[2] It is not designed at this time that you make a detailed

---

[1]Major Julius J. B. Kingsbury, 2nd Infantry, took command of the camp near Sacramento after its former commander, Lieutenant Charles C. Anderson, had died of malaria in San Francisco. The region around Sacramento was infested with mosquitoes, and a malaria epidemic was in full swing by summer 1847. Kingsbury, concerned about his health, took an unauthorized leave during the summer of 1849, faced court-martial, and left the army.

[2]Butte Creek runs for ninety-three miles westward out of the Sierra Nevada, entering the Sacramento River near the present-day city of Colusa, California.

survey of the country that you examine or that your report should be confined to personal observations, on the contrary the General desires that you will embody in your report any reliable information that you may derive from the inhabitants of the country as well in relation to the *General,* as to the *Military* resources of the country.

It is supposed that you obtained at Benicia the instruments etc., that you require for the performance of this duty. If however you were unable to do this you are authorized to purchase such as are indispensably necessary. If without funds you are desired to estimate at your earliest convenience for funds in order that the Commanding General may make the necessary arrangements for supplying you.

You are authorized to employ three assistants at the average wages of the country and to purchase the horses and pack animals that may be necessary. So soon as these duties are completed, you will report in person at Department Head Quarters and the General directs me to impress upon you the importance of completing them at the earliest possible period as your services are greatly required at other points.

These instructions will be sent in triplicate,

Very respectfully Sir,
Your Obdt Servt
Ed. R. S. Canby
Asst Adj Genl

Derby's Journal of a Trip up the Sacramento River September–November 1849[3]

Topographical Officer
Head Quarters 10th Mil Dept.
Monterey, Cal. Dec. 1st 1849

Major

In compliance with the above instructions, which I had the honor to receive on the 17th September ult. on my arrival at Camp Anderson

[3]The original journal is in NARG 393, LR, DP. An edited copy is found at Senate Executive Document 47, 31st Congress, 1st session, part 2. As usual, army editors, likely in California, omitted segments of the journal so that it could be published in the serial set.

near Sutter, Cal I immediately reported to Captain H. Day who I found had relieved Maj Kingsbury in command of the Battalion of the 2nd Infantry stationed at that point.[4] I had not received Department Special Order No. 44, dated Sept. 4th and having therefore no previous information of the duties assigned me, was entirely unprepared without either instruments or funds for their execution.

Finding however that Captain Day was about to move from Camp Anderson, and my services being required in the selection of a site for the new Military Post contemplated, and having no time to communicate with Dep't Head Quartered (as by *ordinary* means of communication a month would be required for that purpose, and an express could not be sent at a less expense than $1000) I concluded to purchase on credit such instruments, animals & other indispensables as might be required for this survey, and to forward to the Dept at once my estimates, trusting that they would be received and honored before my return.

This course I may add was in accordance with the advice of the Senior officers in the station, whose opinion I of course solicited, immediately upon receiving my instructions.

I therefore proceeded at once to engage the services of an assistant, three rodmen, and a teamster, who with my servant and a gentleman named Kemp who volunteered to accompany us as far as Bear Creek composed my party.[5] I had succeeded in procuring a circumferentor,[6] and chain left by Captain Warner[7] (before starting on his late melancholy and fatal expedition), in the hands of Judge Schoolcraft[8] at Sacramento City but was compelled to purchase a Chronometer,

---

[4]Camp Anderson had been named for Lieutenant Charles C. Anderson, who had died of malaria some months before. Captain Hannibal Day, 2nd Infantry, went on to serve in the army during the Civil War, rising to the rank of brevet brigadier general. His brigade held the ground in the wheat field at Gettysburg and was decimated. He retired in 1865.

[5]Kemp's identity is never revealed.

[6]A circumferentor is an instrument used by surveyors for measuring angles, consisting of a flat brass bar with sights at the end and a compass in the middle that is turned over a graduated circle.

[7]Captain William H. Warner, of the Topographical Engineers, had been killed by Indians in an attack in 1849 near Goose Lake, along the Oregon border.

[8]Judge Henry A. Schoolcraft was an early attorney in Sacramento. He represented John Sutter in several cases involving land claims.

Sextant,[9] Artificial Horizon[10] and an additional compass and chain. I then procured from Misters Smith, Bensley & Co.[11] a wagon with six mules for the transportation of our instruments, subsistence stores, forage, etc., the necessary riding animals for the party and such articles of mess furniture etc., as we required, for all of which together with five hundred dollars which they advanced me for the expenses of the road, & they very kindly gave me credit.

I was obliged to pay a much higher price for my mules, and wagon from the fact that the expedition for the relief of the emigrants, under the command of Maj Rucker was at this time being fitted out at Sacramento City, and his large purchases of animals and means of transportation had created a scarcity of both throughout the adjacent country.[12]

We left Sacramento City at 2 P.M. on the 22nd of September, intending to make a short march in advance of the Infantry command, for the purpose of trying our mules, those composing the team having never been worked together in harness.

Leaving Sutter's Fort about half a mile to our right and crossing the bottom land, between the Sacramento and American Rivers, in a direction of N. 30° E. from the city, we arrived at the lower Ford of the American [River], at 5 P.M. having made a distance of nearly two miles in three hours without further accident than breaking the tongue of the wagon, which we repaired upon the spot by wrapping the fractured part tightly with rope.[13]

---

[9]A sextant is an instrument used by navigators for measuring the angular distance between two objects.

[10]An artificial horizon is a horizontal reflecting surface like a mirror or the surface of mercury or other fluid at rest, used in taking altitudes.

[11]Smith, Bensley & Co. was one of the first companies in Sacramento to freight goods into the mines. John Bensley later became involved in developing steamship traffic between San Francisco and Sacramento.

[12]The Donner emigrant party was trapped in late fall 1846 in the Sierra Nevada, some members surviving only as a result of resorting to cannibalism. The army became acutely aware that parties were about to face the same difficulty in late 1849 and assigned Major Daniel H. Rucker to mount relief expeditions. He quickly organized dozens of wagons and supplies and headed east up the trail to meet the emigrants, saving many lives. Rucker later became a general during the Civil War, serving in the Quartermaster's Corps.

[13]John Sutter's fort on the American River was a famous landmark, established in 1839 and improved in 1840 with adobe walls and two headquarter buildings in the center of the compound. See Hurtado, *John Sutter*.

This was owing to the refractory conduct of our team, which upon arriving at the Ford utterly refused to pull the wagon across. So after exhausting the usual arguments, manual and vocal resorted to in such cases, I was fain to hire an individual, who was luckily passing at the time, with eight yoke of oxen, to assist us, which he did with such effect that we soon found ourselves on the opposite bank of the stream with a smooth road before us. The American River, at whose junction with the Sacramento the city is situated, is at this ford about 300 yards in width. The banks are some 30 feet in height but gradual and easy of ascent, which is not the case at the Ford (Chiles') some eight miles above.[14] A sand bar, which at extreme low water is exposed, forming a small island in the middle of the River, makes out from the Southern bank. At the time of the crossing, the water was quite low, varying from 18 inches to 2½ feet in depth but at the commencement of the rainy season it swells rapidly, three days of heavy rain being sufficient to raise it to from four to six feet. The tide rises and falls at Sacramento City, causing a variation in the depths between high & low tides from six to fourteen inches. On crossing the American we passed through a fine grove of oaks which borders the stream through its entire extent, and striking a more Northerly direction arrived without further accident at Dry Creek which at this time was with the exception of a few holes perfectly dry, but which in the rainy season is quite a considerable stream rising in the spurs of the Sierra Nevada and running in nearly a westerly direction to the Sacramento. Its distance at our point of crossing from the Ford of the American, is six miles and a half. Here we encamped for the night finding water in considerable quantity about 1½ miles E. of the road in the bed of the Creek. About 9 P.M. Capt Day's command came up and encamped in our vicinity. The plain between the American River & this creek is of dark alluvial soil, and with the exception of a range of low sand hills running parallel to, at a distance of 6 miles from the Sacramento, is an almost unbroken land extending E. about 20 miles when it rises into low hills which commence the western slope of the great Sierra

[14]Chiles' is a reference to Joseph B. Chiles, who operated ferries across both the Sacramento and American Rivers. Chiles came to California with the first American wagon train in 1841, and later led seven trains from Missouri into the Central Valley.

Nevada. A small field of Tule commences with the Dry Creek extending nearly to the Sacramento.

Upon the commencement of the rainy season the soil upon this plain greedily absorbs the water, and in a few days becomes a thick, tenacious quagmire, which it is difficult, not to say dangerous to attempt crossing, even with pack animals. The Tule at this time is preferable for crossing as its thickly interlaced roots until thoroughly saturated with water, continue elastic, affording for some days a safe passage to Terra Firma.

We made observations for lat [latitude] & long [longitude] by Meridian altitude of the Sun on the 23rd and left our encampment at 5 P.M. proceeding over a plain of precisely the same character as that of yesterday with no alleviation of its unbroken surface to the eye but the distant outline of the timbers bordering the Sacramento and Feather Rivers.[15]

After marching 15 miles N 10° W (by the compass) we came to a small pool of stagnant water where we found several emigrant families encamped, although there was no wood or grass in the vicinity, and proceeding 12 miles farther on came to a small pond, where we encamped at 10 P.M. managing to procure a scanty supply of wood for our camp fires from the stunted alders, which surrounded the muddy & unwholesome looking water. Leaving our encampment at 8 A.M. on the 24th we soon arrived at the Rancho of Nicholas Alteger [Allgeier] situated in a fertile spot at the junction of Bear Creek and Feather River.[16] Mr. Alteger is an old resident of the country and his farm at this place being a well known position has from time immemorial been called "Nicholas' Ranch" he himself being invariably

[15]The Feather River is a major tributary of the Sacramento River, joining it from the east just north of Sacramento.

[16]Nicolaus Allgeier was a German emigrant who occasionally worked for John Sutter. He had apparently married a Native woman who called him "Nicholassee." Sutter and others called Allgeier "Big Nicholas," and his rancho was best known as a place for "grog and cards." Also, visitors often found there a cadre of Native women, of the Nisenan tribe, whom Derby neglected to mention. See Hurtado, *Indian Survival on the California Frontier*, 65, and *John Sutter*, 39, 41, 117, as well as Erwin G. Gudde and Elisabeth K. Gudde, eds., *California Gold Camps: A Geographical and Historical Dictionary of Camps, Towns, and Localities Where Gold Was Found and Mined, Wayside Stations and Trading Centers* (Berkeley: University of California Press, 1975), 243.

known as "Nicholas" only. He has a field enclosed containing about ¼ square miles, apparently of the most fertile soil, and owns quite a large adobe house of two stories in height which presents quite an imposing appearance in this country of log huts & Indian Rancherias. About 100 wretched Indians, playfully termed Christian, live in the vicinity upon the bank of Feather River, subsisting upon acorns (which pulverized with roasted grasshoppers they form into a cake) and salmon, with which delicious fish the river abounds. The more intelligent and docile of these creatures are taken and brought up upon the farm, where in time they become excellent *Vaqueros* or herdsmen & where they are content to remain receiving in return for their services such food & clothing as it may suit the interest or inclination of its owner to bestow upon them. About 1 mile South of Nicholas' Rancho the road divides, the right hand path heading directly across the plain to Johnson's Rancho (now Gillespie's) on Bear creek the left passing "Nicholas' Ranch" and crossing the creek about 1½ miles from its mouth. A path leads from this crossing directly up the bank of the creek and joins the right hand main road again, about six miles south of "Gillespie's."[17]

The right hand main road is the "Truckee route," or Emigrant Trail from the Salt Lake via Truckee River; the left is "Lawson's Route," [Lassen] or the Northern Emigrant trail entering at the head of the Sacramento Valley, near the Head Waters of the Feather River.[18] This latter route is some 300 miles farther than that by Truckee River, but

[17]Johnson's Rancho was developed by William Johnson, who came to California in 1840. It became the headquarters for the rescue effort to save the Donner Party, Johnson marrying one of the young girls who survived the ordeal. It was located near present-day Wheatland, California. Lieutenant Archibald Gillespie carried secret messages from President John Polk, addressed to Commodore John D. Sloat and John C. Frémont in 1846, preparing these men to seize California when news of war arrived with Mexico. Gillespie had several brothers who later became successful merchants in San Francisco. While the Gillespie named by Derby is unclear, it likely is associated with Gillespie & Co., formed by Charles Gillespie and his younger brothers. Derby later identifies the occupant as E. Gillespie. On Archibald Gillespie, see Barbara R. Warner, *The Men of the Fear Flag Revolt and Their Heritage* (Glendale, Calif.: Arthur H. Clark, 1996), 24, 58, 63, 79, 113, 126.

[18]The Truckee emigrant route through the Sierra Nevada starts near Truckee, California, and follows generally the current path of Interstate 80 into the lower Sacramento River valley. Lassen's route originated far to the north, being a cutoff that left Fort Hall, Idaho, into California, through the Warner Mountains (named for Captain Warner), thence in a southwesterly direction, via Pit River, into the Sacramento valley, near Redding.

has the advantage of easier ascents and descents, and generally, affording better pasturage for cattle.

Leaving Nicholas' Rancho we continued upon the river road, passing the crossing, and encamped at a distance of 10 miles from the house, upon Bear creek, in a beautiful grove of oaks & sycamores surrounded by high grass which borders the creek for some two miles in depth on either side and which afforded to our tired animals the most extreme satisfaction. Here, Captain Day concluded to leave his command encamped while making an examination of the country in the vicinity, this point being within five miles of the first [possible military post] proposed in our instructions and within easy riding distance of the remaining points purposed for selection. Accordingly leaving my camp in charge of my assistant (Mr. J. H. Nevelt) I accompanied Capt Day on the morning of the 25th for the purpose of assisting in the selection of the site for the purposed Military Post. Proceeding (by compass) in the direction N 45° E. we arrived at Johnson's Rancho, a small one story adobe building, at a distance of 5 miles from our encampment upon the N. bank of Bear creek. Here we were kindly received and entertained by the proprietor (Mr. E. Gillespie), who volunteered to accompany us in our examination of the Rancho. We accordingly rode over, and made a general reconnaissance of the country in the immediate vicinity, embracing some ten or twelve square miles, and returning late in the evening slept at Mr. Gillespie's house. On the 26th we left "Johnson's Rancho" for an examination of the banks of the "Yubo" River [Yuba] and the adjoining country, striking the Yubo at "Cordua's," or "Specks" bar, direction 25° W distance seventeen miles.[19] Here we found a company of nearly 100 miners busily engaged in the bed of the stream which they had partially exposed by constructing an oblique dam, extending nearly across. They informed us that their operations at this point had proved extremely successful, each individual averaging from two to three ounces per diem. Dear Creek joins the Yubo on the south about two miles above Cordua's and upon the banks of which the richest deposits of gold yet discovered in California have lately been found.

[19]Cordua's Bar was named for Theodore Cordua, a local merchant. Jonas Spect, associated with Derby's "Specks" bar, first found gold on the upper Yuba River. The location is close to present-day Smartville, California. Cordua was actually a native of Germany, and his property now contains the city of Marysville. See Gudde and Gudde, eds., *California Gold Camps,* 82–83.

The result of our examination was the selection of a site about half a mile above the store at "Johnson's Rancho" on the Bear Creek.[20] This point possessed all the merits of others examined and was comparatively free from objectionable circumstances. Being at the foot of the Sierra Nevada slope, the reserve could be made to include a sufficient space of high ground, not subject to the periodical overflows of the creek, building the necessary barracks for the troops, while the rich bottom land bordering the creek on either side afforded excellent opportunity for gardening purposes. A grove of fine oaks answers the purpose of shade for an encampment during the summer months, while the hills in the vicinity are covered with trees, affording a sufficiency of wood for fires, and of logs for houses. The summits of the adjacent hills crop out with a species of sandstone which if quarried would answer admirably for the construction of chimneys, and foundations for buildings. Its proximity to the Rancho is also an advantage, ensuring a constant supply of fresh meat, & we found the water of the creek at all times healthy, cool and pleasant. But the central position of this locality is probably its greatest advantage. The "Truckee Emigrant Route," which was passing an average of 100 wagons and 200 emigrants per diem; the wagon roads to the Yubo mines, that to the "Feather River dry diggings," the trail to Rose's Rancho on the Yubo, striking into "Lawson's Route" at a distance of 20 miles; and the paths to the "Bear Creek diggings" all intersect at this point, while the post is moreover within a few hours ride of all the principle Ranchos and Indian Rancherias or villages in this part of the valley. There are several points on Feather River and one upon the Yubo near Rose's Rancho where much prettier, and more romantic sites may be found, combining most of the advantages of this upon Bear Creek, but the notoriously unhealthy character of these locations, offered an insuperable objection to their being selected.

We found upon inquiry that there had been but little sickness upon Bear Creek during the summer among the Indians, or the emigrants who had encamped for miles upon its banks, while at Sutter's Farm, on Feather River, and at the Ranchos on Yubo, most of the occupants

[20] This was known as Camp (and later Fort) Far West.

had suffered with the periodical fever, and several deaths had ensued.[21] Within three or four miles of the post, gold is found in small quantities in the ravines running towards the creek; digging for which will prove a healthy and profitable recreation for the unemployed soldiers, and will probably be the occasion of preventing many desertions that might otherwise occur.

We returned to our encampment on the 27th of Sept, and made observations for the Lat & Long. The longitude worked out was probably incorrect the chronometer having changed its rate from travelling in the wagon, the Lat we made 39°, 00', 06". We found a very unpleasant incident had occurred on the morning of the 27th. My servant Manuel Montano, who had been with me some two months and had always appeared remarkably honest, faithful & attentive, deserted after cutting a hole in the back of my tent through which he thrust his hand, and removed from a box beneath the head of my bed, my purse containing $435.00 public funds, a gold pencil, gold watch & chain, a pistol and some other articles of value.[22] He also cleaned out his roommates taking whatever money they possessed and left on the best horse we had, carefully selecting the best saddle and riding equipment in the possession of the party. We discovered this at reveille when he had been gone probably some hours, & Mr. Kemp immediately volunteered to start in his pursuit. I accordingly dispatched him with another horse upon the road to Sacramento City, judging that Manuel who was a Chilian [Chilean] and knew nothing of the country would probably take that direction with which alone he was familiar. I have never seen either of them since, (though I have been incidentally informed that Manuel was arrested in San Francisco, confined in a school house, whether as teacher or pupil I did not inquire, and

[21]Malaria epidemics during the summer and fall months had been constant since the Mexican War had commenced. The virus was spread by the large numbers of new people who came to the region, and especially by the new ships that landed at San Diego, Monterey, and San Francisco. While Cook's population estimates are terribly inflated, he is correct in asserting that massive numbers of Native deaths were caused by malaria. See Cook, *The Conflict between the California Indian and White Civilization*; Cook, "Historical Demography," in Heizer, *Handbook of North American Indians*, 8: 92–93.

[22]Derby's experience with hiring men to assist him proved problematic. The man Kemp was sent to find Montano, and neither man returned..See Stewart, *John Phoenix, Esq., the Veritable Squibob*, 61.

had afterwards escaped—But I never have recovered any of the stolen property). This unpleasant occurrence filled us all with surprise and consternation, and threw a general gloom over our little party which it required several days to dispel.

On the 28th, we struck our tents at 8 A.M. and following the course of Bear Creek passing in our progress numerous encampments of emigrants who had halted for the purpose of recruiting their cattle on the fine high grass which abounded on the banks, we soon arrived at "Johnson's" Rancho, passing which about a mile we encamped on the creek nearby at the center of the present reserve. My party having been reduced by Manuel's flight and Kemp's pursuit I replaced them by hiring two persons at the Rancho, the one as cook and laborer the other as an axe man. On the 29th I commenced the survey of the reserve running the base line as nearly parallel to the direction of the stream as possible, which gave it the bearing N 47° E the perpendiculars erected at each extremity having the bearing from the extremities of the base N 43° W. each line one mile in length, and the base 200 yds from and parallel to the South East bank of the creek. This gave the reserve of one square mile, which was situated upon the N.W. bank of Bear Creek, with the exception of the strip 200 yds in width upon the other bank, reserved for the purpose of preventing individuals from squatting or settling in the immediate vicinity of the post. Finding the soil upon the S.E. bank to be equally available for gardens and pasturage, Capt. Day saw fit to include ¼ mile upon the S. E. bank, the lines were accordingly moved this distance to the S. E. preserving however their former bearing. I caused stakes to be driven at every furlong, with a board strongly nailed upon each, and legibly marked "U. S. Reserve." The variation of the compass I found to be 15° 30' East.

We found the heat most intense during the mid day so much so that I was obliged to conduct the work merely during the morning and evening, but in spite of this precaution my assistant and two employees were taken sick, apparently from the effect of exposure to the sun, and being confined to their tents for several days somewhat delayed the completion of the survey. We made observations for the long and lat nearly every day, and found the Latitude 39°, 02', Longitude by the

Chronometer, repeated observations, 120° 45'—but I place but little faith in this, for the unhappy instruments had been so jolted over the rough roads, and had altered its rates in such a very unexpected and inconsistent manner during the march, that we could place but little reliance upon its accuracy. On the 7th & 8th of October, I made an examination of Bear Creek for about 25 miles from the encampment. On the 9th the weather which had gradually, for two or three days been growing damp and uncomfortable changed to rain, pouring down in torrents during the night of the 9th and continuing to rain heavily at intervals during the 10th. Many miners came down from the upper diggings on the Creek during these two days reporting that it was impossible to work on account of the severity of the cold. This sudden change from the excessive heat of the last week, struck me as particularly extraordinary, even in this extraordinary country. On the 14th the weather which had been pleasant from the 10th although much cooler in the mornings and evenings than before the rain, again changed favoring us with a most tremendous gale from the North West accompanied by squalls of rain, which created great havoc among the tents of our encampment, but again all cleared away pleasant before the morning of the 15th. These unexpected rains were somewhat alarming to us, as, having the main part of our work to execute in the examination of the Sacramento valley, we dreaded that they might prove the immediate precursors of the rainy season which if commencing immediately would have put a sudden termination to our labors. On the morning of the 15th having finished the survey, and made a map of the Reservation, which I left with Capt Day for the use of the post, I broke up my encampment at Bear Creek and marched North for the purpose of examining that portion of the valley mentioned in my instructions. I had previously dispensed with the service of my assistant which I had found not entirely necessary, and quite inadequate to the expense of the government. We struck our encampment at daylight but owing to the many little matters of business necessary to be attended to before leaving did not get barely upon the road until 8 AM.

Keeping generally a N by W course we crossed Dry Creek at a distance of three miles from our encampment. This, in winter, becomes quite a considerable stream, which rising in the low hills flows in a S Westerly direction and empties into Bear Creek about 5 miles below "Johnson's." We arrived upon the E bank of Feather River six miles below the junction of the Yubo at 5 PM at a point immediately opposite Sutter's "Hock Farm." Here we encamped for the night.[23]

This is the most beautiful situation that I have seen in California. The river which at this place is about six hundred yards in midst is lined on either bank with majestic sycamores, in a fine grove of which upon the W bank is situated Capt. Sutter's Farm House, a remarkably neat adobe building, whitewashed and surrounded by high and well built walls, enclosing outhouses, corrals, etc. There is about 100 acres of excellent land enclosed and cultivated upon the west bank which yields the most astonishing crops of wheat with very little labor. The river is filled with salmon, and we observed two seines [fishing nets with floats at the top] drawn across about a mile apart from which I was informed the occupants of the farm house obtain a plentiful supply.

About 200 yards above the farm house is situated a Rancheria of Indians, some 300 in number. This village consists of about 20 mud ovens, partly above and partly below the ground with a small hole for egress in the side.[24] They had just commenced the collection of their winters stock of acorns, and had many high baskets containing probably forty or fifty bushels of this species of provender [dry food] standing about. They were mostly naked, and kept up a dismal howling all night as a tribute of respect to one of their number who had departed this life in the day previous. They were suffering much with the prevalent fever, as were the occupants of the farm house, several of whom were sick at this time.[25]

---

[23]Events at Hock Farm, built by John Sutter, can be followed in Hurtado, *John Sutter*. Hock Farm was located on the west bank of the Feather River, just south of Yuba City, California.

[24]Derby's early description of Native populations is interesting in that Albert Hurtado, the biographer of John Sutter, places only sixty-five Indians at the farm. This does suggest a rapid decline from disease, which Derby certainly documents. The Indians located on the farm were Nisenan, who lived in the drainages of the Yuba, Bear, and American Rivers. See Hurtado, *Indian Survival*, 159; Norman L. Wilson and Arlean H. Towne, "Nisenan," in Heizer, ed., *Handbook of North American Indians*, 8: 387–97.

[25]Kroeber estimated Nisenan populations between four thousand and seven thousand people at contact. Derby gives figures that are much smaller in 1849, suggesting considerable decline from malaria. It was afflicting both Indians and whites when he arrived. See Kroeber, *Handbook of California Indians*, 391–95.

Upon the 10th two of my party, Hyes and Flint, were seized with the fever. Having administered a prodigious dose of quinine to each, I proceeded to make an examination of the river in each direction from my encampment which occupied the remainder of the party during the 16th and 17th.[26]

I determined the Latitude of Sutter's Farm 39° 02' 14" nearly identical with that of the Post at Bear Creek. Long by Chronometer 121° 14' 45"—On the 18th one of my patients having recovered, I left this beautiful but unhealthy river, having made a bed in the wagon for Hyes, who although suffering much, refused to be left behind.

We crossed at the lower seine, about one mile below the farm house, having much trouble in getting up the bank on the W. Side, the mules refusing to pull, until by the assistance of some Indians, the baggage was entirely unpacked when by dint of much shouting, screaming and profanity from the party, they managed to draw the empty wagon to the top. Passing the farm house on the left, we now entered upon a fine level prairie, the soil of which was of the richest description and its surface dotted with the "long acorn" oak for a distance of two or three miles from the river. We saw many antelope and deer but did not succeed in obtaining a shot at them. The Buttes were now in full view, presenting the singular spectacle of a range of mountains from 1 [1,000] to 2,000 feet in height, and some twelve miles in length rising from the middle of a broad and level prairie & entirely unconnected with any range, keeping by compass the direction N 45° W.[27] we travelled on until night fall when we had arrived within three miles of the spurs

---

[26]The use of "prodigious doses" of quinine to combat malaria was common by 1849, and within a year, the Office of Indian Affairs was using the drug to treat Indians. The office was also hiring doctors to vaccinate these same Indians for smallpox and treat them for syphilis. The rationale for doing so was not completely altruistic. Medical doctor W. M. Ryer noted that the need for a medical doctor "to attend to the prevention and curation [*sic*] of those contagious and malignant diseases [among the various Indian bands], if permitted to exist among them, would soon spread abroad among the whites." Ryer received two dollars for every Indian treated, most of whom were vaccinated. He claimed to have treated over seven thousand Indians, or virtually all Indians between the Mexican border and the Stanislaus River. See his several reports, Ryer to Agent Adam Johnston, June 1–September 26, 1851, Senate Executive Document 61, 32nd Congress, 1st session, serial 620.

[27]Buttes such as these were often considered sacred by Native Americans. See Peter Nabokov, *Where the Lightning Strikes: The Lives of American Indian Sacred Places* (New York: Penguin, 2006).

of the North Butte [10–15 miles above Sutter's Hock Farm].[28] The heat had been intense, & having crossed the prairie without water, we were suffering extremely for the want of it, as were our animals, whose labor in pulling the wagon over the dry and cracked soil, which yielded at every step, had been very severe.

I therefore caused the party to halt and sent off three men in different directions to search for water, which they found and reported at the Northern extremity of the "Butte" to which place we proceeded at once, and encamped about 11 PM after a very fatiguing days march.[29] The "Buttes" or as they have been generally termed the "three Buttes" have been erroneously represented to be three isolated peaks, rising in the prairie. They are in reality a range, containing some twenty peaks, and about 12 miles in length by 5 or 6 in width. The two principal peaks are at the Northern & southern extremities. The former 2483 feet above the plain, the latter 1841, the Northern "Butte" is much the most remarkable. It is of a nearly conical shape, very steep, its summit cropping out with trap rock disposed in every variety of fanciful figures, one of which (the actual summit) is a tall turret shaped rock, 56 feet high and about 12 ft in diameter. The base runs off into five distinct spurs to the North, between each two of which lies a beautiful green valley, watered by streams having their source in never failing springs about midway up the mountain. The soil is of the richest alluvial deposits, giving no indication of minerals and the whole range is studded with the holly oak, and thickly inhabited by almost every description of game, bear, antelope, black tailed deer, coyotes and even the panther, or puma (called the "Californian Lion") whose roar resounded about our encampment during the night with startling distinctness.[30] We made the Latitude of the N. "Butte" 39° 12' 32" Long 121° 28' 36" of the South, Lat 39° 19' 32", Long 121° 29' 06". It was in one of the Little valleys that we encamped by the side of a beautiful spring of clear ice cold water. Here

[28]Derby here refers to Sutter Buttes, the highest one rising to 2,117 feet, very close to his estimate. They are located ten miles northwest of Yuba City.

[29]The expedition is now ten miles north of Yuba City on the west bank of the Feather River.

[30]The cougars that surrounded Derby's campfire were new to him since they were not generally found in the East. Also called pumas, the "California Lion" actually belongs to the *Felidae* family of cat.

we found many human bones, and the embers of a large fire in which were the remains of a carpet bag or valise & some plates & cups.[31] We observed also a newly made grave in the valley, with a cross placed at its head, on which had been an inscription but it was now illegible. I succeeded in clambering to the summit of the "North Butte" from which I had an uninterrupted view of forty miles in every direction, observing distinctly the courses of the Sacramento & Feather rivers for at least that distance, & on the right and left the Snowy mountains and coast range, extending N & S as far as my eyes could reach.

On the 21st two or three of my men declined proceeding on account of the day, it being Sunday. I had allowed them while engaged on the survey at Bear Creek to rest on this day, but as they were receiving very high wages for very small equivalent, I did not conceive it necessary to cease from travelling, particularly as I very much doubted their sincerity in the matter, I therefore informed them that I had not the slightest objection to halting on every Sunday, but that as they would do nothing on that day, I should infallibly stop their pay on every occasion of this kind—upon hearing this their devotional feelings subsided with vast rapidity and they professed themselves ready to proceed. I had been detained for two days at the Buttes by the sickness of two of my little party, but the bracing air of this delightful spot acted upon them like a charm and by the 22d all were able to move forward. Taking a westerly direction we marched across the plain, which was of rich alluvial soil, but cracked and parched by the heat, until we struck Butte Creek at a point about 12 miles from its junction with the Sacramento, where we encamped.[32] We saw upon the plain while crossing immense quantities of wild cattle, and a large drove of horses containing over two hundred. These animals are I believe perfectly wild and unclaimed, we found it as difficult to approach them as the antelope and elk which in some instances we observed feeding with them. On the 23d and 24th we examined Butte Creek, and the valley beyond, contained between it and the

[31] Such collections of human bones were frequently noted in California, once again likely documenting the horrendous destruction of Indians from disease.

[32] The camp was just east of Princeton, California.

Sacramento. This beautiful stream has its rise in the western spurs of the Snowy mountains, and runs S with little variation in its course some 50 miles where it makes a wide bend to the W in Lat 39° 09' and empties into the Sacramento. It is in many places of inconsiderable width, but everywhere of great depth, carrying I should imagine as much water into the Sacramento as the "Yubo" River, the principal branch of the "Feather" River. Near its mouth it widens to about 600 feet the ground in the vicinity being marshy & covered with Tule & the banks difficult of access on account of the density of the alders and grape vines with which they are lined. There are many clusters of beautiful trees, oaks, sycamores, & ash upon its banks, but it is not thickly wooded as is the case with the Sacramento, & Feather River and its branches. The Plain beyond is of rich alluvial soil, covered with fine grass, which was at this time almost dried up, upon which subsist herds of wild cattle, horses, elk, and antelope. The "tule" swamps do not extend far above Butte Creek, there are but two or three isolated marches of this description on the W bank of the Sacramento. There are two Rancherias of Indians upon this bank, containing probably some 200, male and female.[33] They subsist like those already noticed upon fish & wild grapes in the summer and acorns and pulverized grass hoppers in the winter season, and appeared peaceable enough, but very disgusting to behold being almost without exception stark naked, and exceptively [*sic*] filthy.

Two deputations visited my camp on Butte Creek, they behaved very civilly and received and devoured some biscuits which I administered to them, with a rapacity quite painful to behold. The men were all armed with bows and arrows, & the women furnished with baskets of conical shape. I found out by signs that they were on a grape gathering excursion, probably a sort of barbarous picnic. We found the creek to be plentifully stocked with a large white or grayish fish weighing several pounds, and of a kind that I had never seen before. They resembled bass and were delicious eating.[34]

---

[33]These were northern groups of Maidu Indians. Again, their small numbers suggest considerable demographic decline. For the Maidu, see Francis A. Riddle, "Maidu and Konkow," in Heizer, *Handbook of North American Indians,* 8: 370–86.

[34]According to the California Game and Fish Department, the reference is likely to a striped bass.

On the 25th we marched up the creek about 15 miles and encamped near the crossing of Lawson's [Lassen's] Route. The stream here widens into a little bay about 5 [500] or 600 feet across, very deep, and about a mile in length. We made observations here as we had done at our last encampment (at which we made lat 39° 20' 20", long 121° 41' 15") finding lat 39° 31', long 121° 40' 15". The creek was at this place almost literally covered with ducks and brandt [a "brant" or "brand goose" that breeds in the Arctic and migrates south in the autumn] of which we shot many. We made an observation at sun rise for the variation of the compass which was found to be 16° 49' East—a very large variation but not as great as at the Buttes where we found it nearly 18°. I was disposed to attribute this to local causes or attraction but could find no evidence of iron or in fact any other mineral in the soil. We observed this day as before numerous herds of cattle and horses, one of which came to water at the creek, about sundown within a few hundred yards of our encampment. We observed comparatively few troublesome insects in this part of the country—fleas, so annoying in the low settlements are here almost entirely unknown. Their troublesome office is however disagreeably supplied by an unpleasantly smelling pissmire [the word "pismire" is archaic for ant, and reflective of a peculiar smell coming from an ant hill] which covers the soil, and by its nocturnal rambles effectually banishes sleep. (Solomon's very well intentioned hint, "Go to the ant thou sluggard, consider her ways and be wise," [Proverbs 6:6] is utterly inapplicable to this country, for that industrious insect saves the unhappy sluggard by coming to him with a pertinacity which affords him ample opportunity for a consideration of her ways and being wise enough to avoid making his bed upon the ground.) [parentheses added by army editors at headquarters with the intent of omitting this section][35] We left Butte Creek on the 26th and after travelling S or E about 15 miles, over the dry and parched prairie which was cracked in some places, to the breadth of six or eight inches, rendering the wheeling extremely laborious, we struck "Lawson's Route."

[35]It is clear that army officers were not overly happy with Derby's occasional negative descriptions of people or towns, and they designated key sentences or even parts of paragraphs in his manuscript as material to be "omitted." This occurred on several occasions in this manuscript, and the material was left out of the document when it was published in the serial set.

This we found an extremely good road upon which we marched some 12 miles farther passing many emigrant wagons (filled with dirty and unhappy looking women and unwholesome children) [again, the parentheses in the original journal suggest that this was edited out, as above] and encamped on the bank of the Feather river 6 miles above the ford. The Eastern valley between Butte Creek and the Spurs of the Sierra is from 30 to 40 miles in width of rich soil and covered during the spring and summer with fine grass. It is watered by the Feather River and its branches. This river rises in the Sierra and flows slightly W of south through the valley emptying into the Sacramento. It is remarkably strait throughout its entire course, making few small or abrupt bends and is generally in the same latitudes, wider though not as deep as the Sacramento. Its banks are thickly wooded for some two miles in depth throughout its entire extent, with the "holly" and "long" acorn, oaks, sycamore, beech, ash and alder. Its general depth during the dry season or from the last of May until the 1st of November is from 2 to 10 feet and its bed is much obstructed by sand bars which while rendering fording at numerous points perfectly safe and easy, prevent entirely its navigation even by the smaller class of vessels. It is fed by numerous small creeks, which run down the ravines of the Sierra and in whose dry beds during the summer rich deposits of gold have been and still continue to be found. I observed three Rancherias of Indians upon its bank within 12 miles of the crossing at "Lawson's route," which may contain in all from 3 [300] to 500.[36]

They are all of the same wretched class with those observed upon the Sacramento; they appear perfectly harmless and remarkably good humored, and many of them are in the employment of the emigrants who have squatted in the vicinity. On the 27th we marched six miles down the Feather River crossing in about 2 feet water. A small island in the middle of the stream renders the ford more easy and would be useful in the construction of a bridge at this point, which the high banks render practicable. We kept the river road for seven miles and then crossed the plain to the "Yubo," [Yuba] nine miles farther. This

[36]These northeastern Maidu Indians occupied the north and middle branches of Feather River. See Kroeber, *Handbook of California Indians*, 392; Riddell, "Maidu and Konkow," 370–86.

plain is precisely of the same character as the others unbroken like them save by a few ridges of low hills, and of the same rich alluvial soil. We encamped on the Yubo in the vicinity of Rose's Rancho, a beautiful site for farming or grazing but apparently subjected to but little cultivation.[37] We found here a small adobe house, redolent with the odor of whiskey and festooned with strings of jerked beef, (the occupants of which, one of whom had been in the country for several years, were so wretchedly ignorant, or so profoundly stupified [stupefied] that they did not know the direction to Bear Creek & actually started a doubt as to the existence of such a stream. They spoke of Nye's Rancho at the mouth of the Yubo, as "the house in the corner" and appeared utterly lost to every feeling or sentiment except a vivid appreciation of the merits of the singularly bad whiskey that they were imbibing.) [this section, designated by parentheses, was also omitted from the Serial Set publication]

The "Yubo" is a small but rapid stream flowing S.W. from the mountains and emptying into Feather river. Its bed is rocky giving its water a turbulent character particularly when swollen by the rains or melted snow of the Sierra. Its banks are wooded, and not as high as those of Feather river, and it is occasionally subject to overflow. The soil in the immediate vicinity is of the richest description, and in the upper part of the river, and its tributary "Deer Creek," the richest deposits of gold have been discovered.

Leaving "Rose's Rancho" we marched across the plain to Bear Creek, striking the road from Nye's Rancho to Johnson's at a distance of about six miles from Rose's. We found Capt Day's command comfortably established at Camp "Far West" upon the Reserve and preparations being made for building. We remained here during the 29th and 30th and made observations for the rate of the chronometer, which as I had supposed, had changed its rate very much, from being jolted in the wagon.

The weather had prevented our making but one lunar observation, from which I established the longitude of the Buttes, and my map is

---

[37]Rose Bar Township was created in August 1850, the town of "Rose," or Rose Ranch, being a store built by John Rose a year before. It was located in Township 3. See Thompson and West (no first names), *History of Yuba County, California,* found at http://yubaroots.com; Gudde and Gudde, eds., *California Gold Camps,* 296.

projected from that data. Bear Creek is about forty miles in length, it has two branches which unite about thirty miles from its mouth. Its course is nearly straight as is the case with all the rivers of the valley, and it empties into Feather river, being the second branch of importance in point of size. Its banks are thickly wooded towards its mouth mostly with shrub oak, buck eye and alder.

In the summer it has but little water but is never entirely dry, in the winter it becomes a deep and rapid stream, overflowing its banks to a very considerable extent. I was informed that an instance had occurred of individuals going from "Johnson's Rancho" to Sutter's Fort in a whale boat, the entire plain 45 miles in extent having been submerged during a freshet, but I do not vouch for the truth of the statement. Leaving camp Far West on the 31st we travelled down the creek to its mouth then continuing upon the bank of the Feather River until we had arrived within a mile of the town of Vernon [Verona], situated at its mouth and supposed to be the head of navigation upon the Sacramento, where we encamped. Feather River near its mouth is a very broad and beautiful stream. Its banks are heavily timbered and some 50 feet in height, coming down abruptly to the water. There is a sufficient depth of water as far as the mouth of Bear Creek to float any small sized vessel, but the frequent occurrences of extensive sand bars renders the navigation to this point at present impracticable.

During the night we killed three veritable raccoons, the first that I had ever seen in this country. The town of Vernon [Verona] is situated at the junction of the Feather & Sacramento rivers and that of Fremont immediately opposite. Each contained some 20 houses and one or two hundred inhabitants. (The speculators in land whom we found very busy about these two little places assured me that they would yet rival New York and Brooklyn in size, business and population, an opinion which I did by no means attempt to contravene.) [this sentence omitted from Serial Set version]

The valley of the Sacramento on the Western bank of that stream is for the most part a barren plain, with little vegetation or water, it is from 30 to 45 miles in extent being bounded on the West by the Coast Range of mountains; there are no streams emptying into the Sacramento from the West, south of the latitude 39° N with the exception of "Puta" and "Cache" creeks. The latter is an outlet of Clear Lake, flowing from its S.E. extremity and losing itself in the "Tule" swamp which borders the western bank of the Sacramento, about six miles S.W. of the mouth of Feather river.[38] Puta Creek rises in the Coast Range and flowing S.E. empties into the Sacramento about 15 miles below the mouth of the American River.[39] The whole country between these creeks is liable to overflow, and is very dangerous to attempt travelling after a heavy rain. The "tula" swamp upon the western bank of the Sacramento, extending to the vicinity of "Butte" Creek and occurring occasionally above, is from three to six miles in width and is impassable for six months out of the year. A corduroy road [a road made of logs laid crosswise] may however be made over it, which has been used in some instances with success at all seasons. There is a short road of this kind in the rear of Sutter which I believe has proved entirely successful. We reached Sacramento City on our return upon the 2d of November and encamped upon the outskirts during a tremendous storm of wind and rain, which proved to be the bona fide commencement of the rainy season. Having paid and discharged my party with the exception of the teamster and Mr. John Day whom I retained as an assistant, and finding it impossible to sell my animals and wagon to advantage, in consequence of the identical reason that prevented my purchasing (i.e. the expedition for the relief of the suffering emigrants, which had just preceded my return and whose mules and other property were being disposed of at auction) I concluded to take them all to "Pueblo de San Jose" where I had reason to believe they would meet with a ready sale.

I accordingly crossed the Sacramento on the 6th at 2 PM and encamped on the verge of the Tule being unwilling to run the risk of remaining all night upon the plain beyond. It rained tremendously all night and we were glad to take the road at daylight in the morning. We crossed the "Tula" safely but found the road beyond extremely

38Cache Creek empties out of Clear Lake, which is some fifty miles north of Napa, California.

39Puta Creek was not well known at the time, its main course running southeasterly, emptying out of Berryessa Lake, some twenty miles north of Napa.

heavy and covered in some places with water. There we passed without difficulty observing to keep carefully in the road, which being packed by travel was not as dangerous as the plain upon each side. We had arrived within half a mile of "Puta" Creek when I observed with astonishment and alarm that a strong current was setting down the road & that the water was deepening around us with rapidity.[40] I at once comprehended that the creek had overflown [*sic*] its banks and that we were in a dangerous position. I gave the order to the teamster to turn about immediately, but it was too late. The mules sank at once on turning from the road, the wagon was fast locked in the yielding mud, and the water as we afterwards found was gaining on us at the rate of four feet an hour. It was with the utmost exertion and no little danger that we succeeded in getting the mules extricated from the wagon from which I had already saved my chronometer and best sextant, my drawing instruments and papers. We were compelled to abandon the wagon, with the remainder of the instruments, and all our personal property and return to Sacramento City, where I disposed of my animals at auction. The whole of the valley I found on my return had been rendered a perfect quagmire by the recent rains and several wagons had been lost on the same day in an attempt to go to Fremont from Sacramento City. I arrived at Benicia on the 10th of November ult [ultimate], from which place I had the honor to send you a detailed report of the loss of my wagon and other circumstances connected with my expedition.[41] ~~I have the honor to enclose with this report a topographical map and copy.~~

From the above general summary of the journal kept while engaged upon the expedition, and from the accompanying map, an idea may be obtained of the geographical position of that portion of the valley mentioned in my instructions, and as far as the means of my communication, its capacity of supplying forage, & its agricultural character are concerned, of its *general,* and *military* resources. It will be seen that there is no point in this portion of the valley at which a military

[40]The route traveled is roughly equivalent to Interstate 80, between Sacramento and Vallejo, California.

[41]Benicia, on Suisun Bay, was the supply depot for the army in California at the time.

post can be by any possibility be required, that is absolutely free from objection. That, [those positions] already selected perhaps combines as many advantages in a military point of view, as it is possible to obtain. A position at the North Butte or upon Butte Creek at some point near the crossing of Lawson's [Lassen's] route, would be undoubtedly more pleasant in every respect, perfectly healthy, and would possess every requisite for the comfortable subsistence and shelter of the troops, but beyond the attaining of these points I can conceive of no advantage to be derived from establishing a post in that vicinity.

There would be no inhabitants to protect, and nothing in fact to protect them from (and unless engaged in War of extermination upon the "coyotes" the lives of the garrison would pass away in inglorious idleness). [this section, in parentheses, omitted from Serial Set version] As far as regards *building materials* all points are equally eligible, the soil of any portion of the valley mixed with water and chopped grass, and exposed to the sun makes excellent "adobe" or sun dried brick, which is probably more economical, easier of working and better adapted to the climate than either timber or stone. Prime timber may be obtained however, of the best quality upon the table land, at a distance of about 20 miles from the commencement of the hills, the oak, which forms the greater portion of the timber upon the streams, being too hard and brittle to work easily, and soon decaying when exposed to the weather. Two steam sawmills are now in process of erection upon Bear Creek, about 4 miles above the newly established post from which when in operation sufficient supplies of Pine timber may be obtained.

There are two Rancherias of Indians upon the Sacramento, one upon Butte Creek, three upon Feather River and one upon Bear Creek (about 10 miles above the post) which have come under my observation. All of these together must contain something under one thousand individuals, including men, women, and children.[42]

---

[42]These were all part of the Maidu tribe, and once again Derby is giving a population figure for the Indians of the entire lower Sacramento River drainage as a mere one thousand people. Later evidence will show that the numbers in the mountains to the east were not any larger, and may have been smaller.

I was informed that upon the upper Feather and Yubo rivers, were some two or three thousand living in the hills, but whether belonging to the same tribe as those of the valley I could not ascertain.[43] All that I have seen appear equally ugly, harmless, & inoffensive but being perfectly barbarous and acting as I imagine more from instinct than reason, they are liable to commit at any moment some unexpected outrage for which neither themselves nor any one else can assign a reason. I was informed that the Rancheria upon the Sacramento had within a few months committed three murders upon white men traveling upon the western bank. I was not able however to arrive at any details and am not satisfied that the report was worth of credence.

As far as I could ascertain by inquiry, from those persons most likely to be best acquainted with their character and habits all of these Indians are to be viewed with suspicion and distrust, and I found it generally conceded that those termed "Christian Indians" who by intercourse with the whites had added to their original qualification of low cunning, some gleams of intelligence, were by far the most dangerous, being invariably found to be the ringleaders in all thefts or other outrages committed by a Rancheria.[44]

Should the present rapid emigration to this country continue during the ensuing year, the entire valley will undoubtedly be thickly settled, with a hardy population, who attracted by the fertile soil and beautiful scenery of the banks of Feather River and its branches, will brave its sickly climate preferring to reap a sure and lucrative harvest from agricultural pursuits, to enduring the hardship, exposure, and sickness of the mines for a doubtful prospect of immediate wealth. In this case the post at Bear Creek instead of being as now, on the frontier of civilization would be surrounded by a population perfectly able and willing to take care of themselves, and it might become advisable

[43]These were Maidu Indians, the same as those in the valley. Derby's figures, gleaned from questioning individuals and observations made at the various villages he visited, suggest a total population of just four thousand Indians who inhabited the lands north of the American River, from the Sacramento valley into the eastern slopes of the Sierra Nevada range. Later expeditions, published in chapter 4, demonstrated that most of the Pit River region, extending into Goose Lake, a mountainous region, remained completely unoccupied by Indians.

[44]This was a general view, that former mission Indians often became ringleaders for stealing parties.

to establish a station either farther to the North in the valley or in advance upon the "Truckee" road, according as either became finally the main route for emigration. I have the honor to enclose with this memoir a map and copy of the Reserve at Bear Creek, also a map and copy of the Sacramento valley ~~to the mouth of Butte Creek~~ from the American River to the mouth of Butte Creek, which embraces that portion of the country referred to in my instructions.

Trusting that they may meet the approval of the Commanding General, I have the honor to remain, sir

With high respect
Your Obdt Servt
Geo H. Derby
Lieut Top Engineers

Maj. E. R. S. Canby
Adj Genl 10th Mil Dept.

"Reconnaissance of the Tulares Valley . . . by Lieut. G. H. Derby, Topographical Engineers, April and May 1850."
From Derby's survey of the Tulare Region
(the southern end of the Central Valley), April–May 1850.
Courtesy of National Archives and Records Administration.

CHAPTER 3

Derby's Survey of the Tulares Region 1850

THE EFFORTS OF THE ARMY TO LEARN THE GEOGRAPHY OF California had reached the point by 1850 that more isolated regions of the state increasingly attracted the attention of senior officers. Given their isolation, the army knew little about the extent of Native populations. Most of the Indians of these locales had only occasional contact with Europeans, and some had fled Spanish missions, becoming leaders of raiding parties. Newspaper editors believed that many of these regions had extremely large populations of Indians. If so, they would be important in determining army policy relative to the allocation of the scarce number of troops then allotted to California. Lieutenant Derby's next assignment, then, included a close examination of the Tulares region of southern California, the lakes to the south of the San Joaquin River valley, and the various tributaries that fed them. These lakes were surrounded by tule, or tall reeds, that made the region difficult to travel in, especially in the summer. While Spanish expeditions had explored these regions in years past, they left virtually no maps or information about them. It was up to the topographical engineers, like Derby, to fill in these blank spaces in the geography and human population of the southernmost portion of the Imperial Valley.

Derby's journal of the Tulares region, included hereafter, is preceded by Major Canby's orders to him.

Orders to Lieutenant Derby

Headquarters, 10th Military Department[1]
Monterey, California, April 9, 1850

Sir:

The objects of the reconnaissance which you are about to make have already been verbally indicated to you, and I am now instructed by the commanding general to direct your attention again to the most important of these objects.

1st. An examination of the country between the latitude of San Miguel and that of San Luis Obispo, for the purpose of selecting a position that will cover the passes leading from the Tulares into the settled country east of San Luis Obispo.[2]

2nd. An examination of the passes between San Miguel and Santa Margarita, and of the valley of the Tulares from that neighborhood to the King's river, for the purpose of ascertaining whether there is a practicable route for loaded wagons from the coast to King's river.[3]

This route should cross the Sanjon de San José at or near its junction with the lake; and it is important to ascertain whether any portion of this route is overflown in the seasons of high water.[4] If so, what portion of it and for what period of the year. The route from San Luis Obispo has already been examined by Captain Warner, and it will only be

---

[1]Canby's orders to Derby, April 9, 1850, NARG 393, LR, DP.

[2]San Miguel Arcángel Mission was constructed in 1797, some forty miles due north of San Luis Obispo. It had a Native population of just under a thousand Indians ten years later but, like all other missions, was secularized in 1835.

[3]Santa Margarita was a rancho by the 1830s, originally founded to assist in the collection of Indians for the San Luis Obispo mission. It was located eight miles north of the mission.

[4]The Sanjon de San Jose refers to a large swampy region, about forty miles in length, that connects Lake Tache with the San Joaquin River.

necessary to examine that portion of it that lies east of the San Luis and Monterey road.[5]

If, however, you should have time, it will be well to ascertain the present condition of the road from San Luis to Santa Margarita, and also whether the land transportation may not be diminished by landing supplies at Estero Bay, instead of the Bay of San Luis. Captain John Wilson is suggested to you as a person from whom you may obtain much reliable information in relation to this subject as well also as to the safety of Estero bay for vessels. If you should visit Estero bay, the latitude and longitude of the Moro will be determined.[6]

3rd. The examination of the country on the eastern side of the Tulares from the San Joaquin river on the north to the latitude of ——— on the south, for the purpose of selecting a suitable position for the establishment of a military post in the neighborhood of the Tulare lakes. [Tulares usually referred to the region, whereas the lake was called Lake Tache (Laguna de Tache).]

The objects to be attained in making this establishment will be the protection of the frontier east of San Luis Obispo from Indian incursions, and the control of the Indians inhabiting the borders of the lakes and the slopes of the Sierra Nevada, east and north of those lakes. The No-tonto village occupies a central position for these purposes, and has heretofore been suggested as a suitable position for the establishment of a military post.[7] You will give a particular attention to the examination of points in that neighborhood, indicating the several points that may possess the necessary requisites for military posts, and reporting particularly for the information of the brigadier-general commanding; the comparative advantages of each; the military and general resources of the country in the neighborhood; the facility with which lateral communication may be made for the purpose

---

[5]This road was designated as the Camino Real, or King's Highway, but some early maps simply call it the Salinas–Monterey Road.

[6]Estero Bay was ten miles north of San Luis Obispo, now called Morro Bay.

[7]The Notónoto village, on the San Joaquin River, was first identified by the Spanish. See Kroeber, *Handbook of California Indians,* 484, 491; William J. Wallace, "Southern Valley Yokuts," in Heizer, *Handbook of North American Indians,* 8: 460–61. The lake is usually referred to in early documents as Lake "Tache," while the Indians living along its shores were called the "Tachi."

of operating against the Indians of that country, and the nature and length of the route by which the command to be stationed there must be supplied. Particular attention will also be given to the collection of such information as may be useful in guarding against the selection of an unhealthy position; and among other facts, you will ascertain, if practicable, the direction of the prevalent winds during the summer and fall months, and the nature of the country (whether marsh or otherwise) over which they pass.

4th. The selection of the route by which supplies are to be sent from the coast being of very great importance, you will bestow particular care in the examination of such as may be discovered by or suggested to you. In addition to that already indicated to you, there are two others to which your attention will be directed. The first of these is from the highest point of navigation on the San Joaquin by the valley of that river to the neighborhood of the point at which the post is to be established. The second is from Monterey by some practicable pass in the coast range, if one can be discovered, to the neighborhood of the same point.

Both of these routes will probably cross the Sanjon de San José at its junction with the San Joaquin river, and be the same for the remainder of the distance. For the first of these routes it will only be necessary to ascertain the point to which vessels may ascend the San Joaquin, and the length and nature of the road from that point to the crossing of the San Joaquin or the Sanjon, as the case may be; and for the second examination of such of the passes in the coast range between Pancheca's and Miguel as may promise to be practicable for loaded wagons.[8]

Care will be taken to select for the passage of any streams that it may be necessary to cross, points at which the banks on both sides are accessible by wagons at all seasons of the year, that the length of the different routes should be ascertained, and that the nature of any obstacles should be fully reported.

[8]Canby is referring to San Miguel in the south, and the pass named after it that led into the southern end of Tache Lake, and Panoche Pass, in the north, which was an extension of the pass leading west from San Juan Bautista. The Panoche Pass crosses into the San Joaquin basin east of present-day Coalinga, California. See Gudde and Bright, *California Place Names*, 237.

In addition to the above you are desired to collect and report, for the information of the commanding general, any reliable information in relation to the general and military resources of the country through which you pass; its geological structure, &c., &c.; and to determine as accurately as your means will permit, the latitude and longitude of the northern extremities of the Lakes Tache and Buena Vista, and any other important points.[9]

The infantry company under the command of Lieutenant Moore is designed to cover your operations while engaged in the Indian country in the vicinity of the Tulares, and that officer will be instructed to do this effectually, and to render you any assistance that may be needed.[10] For the ordinary purposes of assistance and protection in the performance of your duties, the topographical party heretofore authorized (one guide and eight men) will be sufficient, if your operations are conducted within a reasonable distance of that command. If, however, that force at any time be found insufficient, you will make requisition upon Lieutenant Moore for any escort that may be deemed necessary.

The commanding General does not anticipate any difficulties with the Indians that you may encounter in your operations, but it will be necessary to be constantly on the guard, and that you should not at any time place yourself beyond the reach of assistance. You will see that the Indians whom you meet are treated kindly by your party, but you will not suffer them to remain in large parties about your camp. The utmost economy will be observed in your operations; and the commanding general directs that no expenditures be made, or obligations contracted, except such as are indispensably necessary.

Very respectfully your obedient servant,
Ed. R. S. Canby
Assistant Adjutant General

Lt. G. H. Derby, T.E.
Monterey, Cal.

---

[9]Buena Vista Lake was at the southern extreme of the Tulares region, near present-day Bakersfield, California.

[10]Canby is referring to Second Lieutenant Treadwell Moore, who later commanded at Fort Miller.

Derby's Survey of the Tulares Region April–May 1850

Geo. H. Derby[11]
Lieut Topl Engineer
Report on The Tulares Valley etc.
April and May 1850

Monterey, July 10th 1850

Major

I have the honor to report that in compliance with the above instructions I left Monterey upon the tenth of April with my party and the escort assigned for the duty and proceeded at once to completion of the duties assigned to me in articles 1 & 2 which occupied me until the 26th when I returned to San Miguel and reported to you in detail the result of my operations. In the discharge of these duties I measured and took the bearings of the road from Monterey to San Luis Obispo, observing also the latitude and longitude of the principal intermediate points (a summary of which will be found in the margin) & I examined thoroughly the ranges of the Coast Mountains to the East of the road for the purpose of ascertaining if there could be any communication with the sea from a point between San Miguel and San Luis Obispo.

The road from San Miguel to San Luis Obispo I found generally good, between "Passo des Roblas" [Paso Robles today] (a rancho upon the road about 15 miles from San Miguel and Santa Margarita) there are two or three places however where for a short distance the ground is constantly marshy, and there is a long and somewhat steep hill to be surmounted three miles W. of San Luis Obispo but these are difficulties of no particular importance and the road may be and is in fact at present constantly travelled by vehicles of every description.[12] The bay

[11]Derby's original journal is found at NARG 393, LR, DP. An edited version was published in Senate Executive Document 110, 32nd Congress, 1st session.

[12]Santa Margarita, Paso Robles, and San Miguel Mission are all on the road leading north from San Luis Obispo. The Rancho Paso Roblas was given in a 26,000-acre Mexican land grant to José Pedro Narváez in 1844. Mission San Miguel Arcángel was built in 1797 by Franciscans. Heavily damaged in the 2003 San Simeon earthquake, it has just recently been reopened to the public. See Rose Marie Beebe and Robert M. Senkewicz, *Testimonios: Early California Through the Eyes of Women, 1815–1848* (Berkeley: Heyday Books, 2006), 85–93.

of San Simeon in Lat 35.40 N is in appearance a good roadstead and is well protected from the N and W but I was informed that both this and Esters bay which is an indentation in the coast in Lat 35° 20' N. (9 miles N of San Luis Obispo) are impracticable for anything but the smaller class of vessels having but about 8 or 10 feet of water and their entrances dangerous and obstructed by rocks and sand bars. The Bay of San Luis Obispo on the contrary I found highly recommended as easy of entrance safe and commodious but as I was obliged to obtain my information from residents of that place which with its few half ruined adobe buildings they persist in considering a thriving city. I think it possible their opinion on the subject may be slightly prejudiced. The bay appears well however and is certainly sufficiently extensive. I observed a large brig lying at anchor inside, while there.

As I had no instructions to that effect and it would have occupied some time I made no particular examination of these bays or the intermediate coast but I would respectfully suggest the propriety of having a general survey made of the Bay of San Simeon as several opinions to which I listened on the subject I found materially conflicting.

Between the valley of Santa Margarita and the coast, a distance of about 21 miles there are two distinct ranges of mountains which we examined N. & S. for 15 miles. They are steep rocky and barren or covered with shrubs, and no pass can be found through or over them to the coast S. of Santa Margarita.

From "Passo de Roblas," however I discovered a horse trail running almost due W. which although it crosses two very declivitous hills and is seldom travelled may I think be made a wagon road to San Simeon Bay (should that be found of importance). At present the residents of the ranchos in the vicinity of San Simeon and Estero Bays in travelling to Monterey take the beach, around by San Luis Obispo, and they appeared much surprised that we had succeeded in getting our mules over the mountains at all. We found but three passes through the Coast range to the west of the road between the latitude of San Miguel and that of San Luis. These are the two roads the one from San Miguel the other from San Luis (via Passo de Roblas) which meeting at a point called Estrella from the Pass of San Miguel and a road passing through a Cañada in the hills about 15 miles E. of San Luis called the "Panoche pass" which debouches near the head of Buenavista Lake and might with a little labor be made an excellent wagon path.[13] This last in the event of San Luis becoming a large sea port & the probable discovery of rich mines upon the streams of the Tulares Valley may become of great importance. Having finished my examinations and obtained what information I could with regard to the Tulares Valley I returned with my party to the Infantry encampment upon the 26th.

We left San Miguel on the 28th and crossing the river within 300 yards of the Mission proceeded up a deep ravine in the hills through which we found a trail leading out upon the banks of a small stream at a distance of two miles from the river. This stream has its rise near the summit of this portion of the range, it is about 18 miles in length and flowing in a S.W. direction empties into the Monterey river about a mile above the mission.[14] With the exception of two or three springs in its bed it is dry during the summer and its valley forms the Western portion of the pass. Keeping the bank of the stream we passed through a beautiful and fertile valley, over an excellent road arriving at Estrella early in the afternoon. The distance of this point from San Miguel we found by the viameter to be 12.77 miles, it is a beautiful spot covered with fine large oaks and with a little cultivation the soil might be rendered extremely productive.[15] There is a never failing spring of water near the bank of the creek which we discovered and enlarged. Four valleys diverge from this point through the S. West, one of which leads to the road from San Luis Obispo entering from the main coast road near the Rancho of "Passo de los Roblas." The peculiarity of the divergence of these four valleys and their corresponding ridges from this point resembling the rays of a Star has given it its

[13]There are two possible routes eastward from San Luis Obispo, one passing just north of Pozo Summit, with an elevation of just 2,635 feet, and one to the north of it, which constitutes California Highway 58 today. Derby is unclear as to which one he is referring to. His reference to "Panoche Pass" is likely incorrect, as it follows County Route J1 into the San Joaquin valley, farther north. However, *Panoche* is a common word, deriving from places where Indians collected syrup from plants.

[14]By Monterey River, Derby in fact means the Salinas River.

[15]A viameter counts the revolutions of a wagon's wheels to calculate distance.

very appropriate name Estrella. We encamped here on the 28th and leaving early the next morning proceeded through the pass having previously sent forward a party of pioneers to make any repairs upon the road that might be found necessary. Keeping the valley of the stream for about six miles bearing about E. by N. we struck a narrow valley or cañada through which we passed and emerged after travelling about 11 miles upon a plain some six miles in length by one or two in width crossing this plain and still continuing to follow the course of the stream we encamped about 2 P.M. at its source a point about 2 miles from the summit of the range and distant 16.83 miles by the viameter from Estrella. The road we found excellent no repairs being required the country barren but little grass and a few stunted trees and shrubs growing on the margin of the stream. We observed many antelope feeding in the cañadas upon our right and left and killed a large rattlesnake while on the road. Near our encampment three distinct ridges of trap rock about half a mile apart run through the hills as far as the eye can reach presenting precisely the appearance of the foundation of ruined walls. They vary in height from six inches to twenty feet and are about six feet in thickness, but the most singular circumstance connected with them is the fact that they lie as nearly as possible in the plane of the true meridian. We have found at San Miguel by observation the magnetic variation to be 15° & 28' E. and we now found the bearing of these ridges by the compass N 15° 30' W. On the 30th we crossed the dividing ridge without difficulty and arriving at a spring about five miles from the summit concluded to encamp although we had made but 6.¾ miles by the viameter as I found we should find no more water until our arrival at the lake which I thought as too great a distance for one days march.

The road we found very tolerable but requiring some little repair as the trail lay upon the sides of the hills which in some places were quite steep. Upon the whole I have no hesitation in saying that the road through the pass is quite as good and needs as little repair as any portion of the road between Monterey and San Miguel of the same length, and is by far the best pass through the coast range of mountains not excepting Livermore's Crossing a plain 6.04 miles by the viameter in width.[16] Upon the 1st of May we came to a small stream called *Dick's* Creek which rising in the hills flows South for about 12 miles and loses itself in the sandy soil forming during the rainy season quite an extensive marsh at its termination.[17] Here we found good water, and the banks of the creek lined with wild oats affording excellent grazing for the animals. I concluded therefore to encamp here and proceed with my party to examine the trail leading in the direction of the lake which I hoped to be able to reach and return from before sun down. The escort having arrived and encamped in our vicinity I started accordingly at about 10 A.M. and crossing two ranges of low hills on a broad and smooth trail arrived on the shore of the great Tache Lake about 1 P.M.[18] We were unable to get close to the water in consequence of the Tulé which environed it extending into the Lake from 200 yards to ¼ of a mile as far as the eye could reach. With a glass I could distinguish the timber at the North and the Tulé at the South end of the Lake the length of which I estimated at about 20 miles, but we could not distinctly make out the opposite or Eastern shore. The peaks of the Sierra Nevada at this place 12000 feet above the level of the sea and covered with perpetual snow appeared in close proximity and rising far above the horizon seemed to us to come down precipitously to the very edge of the water. The distance from our encampment to the Lake we estimated at 18 miles or nearly a days march and as the country passed over was a perfect desert and I found here no forage for the animals but wire grass, the water standing in the Tulé Marshes brackish & no wood at all I concluded to return immediately to camp and in the morning to make a reconnaissance to the South of our position for the purpose of finding a road to the

[16]Livermore's Crossing was well known by this time as a road leading over the hills between San Francisco Bay and Stockton.

[17]Dick's Creek, which Derby refers to, apparently drained eastward into Tache Lake.

[18]Tache Lake, or Laguna de Tache, sometimes referred to as Tulare Lake, was the largest in the chain of lakes and swamps generally making up the Tulares region. It was fed by the Kern, Tule, and Kings Rivers, coming out of the Sierra Nevada. When farmers began diverting water from these rivers for irrigation in the late nineteenth century, the lake dried up. It was likely at one time the largest freshwater lake west of the Mississippi River. Lake Tahoe has that distinction today.

Southern extremity of the Lake which point I hoped to be able to reach in one days march.[19]

An examination was accordingly made on the 2nd (a portion of the party being left in camp to cut wild oats which I purposed to transport for forage, as it was evident we would find none upon the shores of the lake) which terminated favorably a good path being found through the Southern extremity of the valley, and a trail leading apparently around the South of the lake. On the 3rd we broke up our pleasant encampment at Dick's Creek and succeeded in reaching the Southern part of the lake, where we encamped upon the sand for the night having marched 24.29 miles nearly in an Easterly direction from the termination of the pass. We found here a ridge of sand about 100 yards in width and 12 feet above the level of the lake which divides the water of the Northern or Taché from the bed (now nearly dry) of the Southern or Ton Taché Lake.

This last is little more than a very extensive swamp covering the plain for 15 miles in a Southerly direction and is about 10 in width. It is filled with sloughs and small tulé lakes and is of course impassible [*sic*] except with the assistance of boats or rafts. The gradual receding of the water is distinctly marked by the ridges of decayed tulé upon its shore and I was informed and see no reason to disbelieve that ten years ago it was nearly as extensive a sheet of water as this Northern Lake having been gradually drained by the connecting slough and its bed filled by the encroachment of the Tule.[20] We crossed the slough on the 4th being assisted with rafts of Tule by the Rancheria of *Ton Taché* Indians which we found established at this points. They were about 100 in number mostly clothed and very friendly. The captain was an old Indian from the San Luis Obispo Mission and spoke Spanish indifferently well.[21] Several of these Indians had bits of paper on which were written recommendations signed by

[19] Wire grass is actually native to the southeastern United States. Derby was likely referring to a species of similar makeup.

[20] Derby seems to be suggesting a considerable, and long-lasting, drought.

[21] The Spanish called these Indians Tulareños, being residents of Los Tulares district. They were of the Yokuts family, Indians that dominated the San Joaquin River basin and the various tributaries, such as the Kern and Kings Rivers, that emptied into it. See Kroeber, *Handbook of California Indians,* 474–76; Wallace, "Southern Valley Yokuts," 448–61.

various persons, Spanish and American by which they set great store and eagerly exhibited for our examination. I gave the captain a certificate that he had treated us kindly and proceeding on over the desert which we found very painful travelling for the animals, encamped on the S.E. point of the lake having made 13.64 miles.

We found here another small Rancheria called the *Tente-Tachés,* living like the others principally on fish and reptiles and numbering about 50. I had a conversation with the Captain who was quite an intelligent old fellow originally belonging to the mission of San Miguel. He informed me that the Taché Indians of which tribe his Rancheria formed a portion numbered about 800 in all and were settled on the shores of the great lake but their principal Rancheria containing about 300 is situated at its N.W. extremity.[22] They are all peaceable and friendly in their dispositions and he assured me that they had never stolen or eaten any horse flesh, which judging from the meager condition of himself and companions, I should think highly probable. We gave them some bread and a little sugar with which they were highly gratified and in the morning brought us some dried fish in return which we accepted but could make no use of.

On the 5th after marching 15.38 miles over a continuation of the barren, sandy desert which had been our route for the last three days, we came to two or three cottonwood trees upon a small stream called Moore's creek, which empties near this point into the Ton Taché swamp and here the grass being very tolerable we encamped.[23] This is the first point on our route where good water, grass or wood is to be found after leaving Dick's Creek, it will therefore be seen that it is necessary for loaded wagons to transport if travelling this route sufficient grain to forage their teams for three days. As I deemed it unnecessary for the Infantry escort to accompany me through the desert to Buenavista Lake, which it now became necessary to examine, I proceeded up Moore's creek on the 6th for 12.20 miles to a point about 5 miles W of the high peaks of

[22]While also Yokuts Indians, the term "Tente" in unexplainable, other than perhaps being a snide way that Derby used to describe their eagerness to touch, or grope, people, coming from the Spanish verb *tentar.*

[23]As a result of irrigation in southern California, Creeks such as Dick's Creek and Moore's Creek likely have simply disappeared. Others mentioned in his journal have no doubt suffered a similar fate.

the Sierra Nevada where I found an excellent encampment, the creek at this point and above among the hills being lined with cottonwood and willows, and some large oaks, the grass among which was growing luxuriously. Here Lieutenant Moore encamped his party to await my return from the Buenavista Lake, an examination of the country in the vicinity of which I thought might take five or six days.[24] On the 7th taking my party I ascended Moore's creek for about 20 miles among the hills, and found it a small rapid stream about 50 feet wide, the water extremely cold from the melting of the snow upon the Sierra and about two feet in depth. We discovered two small branches upon each of which we observed the ruined remains of a large Rancheria.[25] The banks of the stream as well as the hills surrounding were heavily timbered with oaks and three large species of pine.

Leaving Moore's creek we crossed through the ravines in the mountains to Tule river, which runs in the same direction as the former stream (a little N. of E.) and at about 8 or 10 miles N. of it, emptying into the S. Eastern extremity of the Taché Lake. This stream has two branches, the upper portions of which are well timbered, but the banks are swampy near the lake, and for a long distance on the plain, the Tule running up to within 5 miles of the hills. At this time the stream was about 100 yards wide, from 12 to 20 feet deep and very rapid which last is a general characteristic of all the streams to the East of the lakes. Upon its upper waters and their vicinity in the hills plenty of large pines are found, lower down it is well timbered with the different species of oak, sycamore, cottonwood and willow. Ascending the stream about 10 miles we suddenly came upon a Rancheria of Indians in a sequestered nook of the hills. We swam the river and were met upon the bank by all the men (some 60 or 70) belonging to the band. They received us favorably although with evident distrust and informed us that they belonged to the *Thulime* [Tulamni] Tribe.[26] I had been previously told

---

[24]While Derby does not say so, likely he named Moore's Creek in honor of the young Lieutenant Treadwell Moore, who commanded the infantry escort. The name since has been lost to history.

[25]Derby mentions several abandoned Indian rancherias in his journals, no doubt indicative of the impact of disease on Indian bands.

[26]These were actually Tulamni Yokuts, whose general habitat was the region east of Buena Vista Lake. See Kroeber, *Handbook of California Indians*, 478; Wallace, "Southern Valley Yokuts," 448, 461.

by the Captain of the Ton Tachés that they were a hostile, thieving nation, and observed about their Rancheria several horses, and skeletons of animals, the appearance of which was calculated to corroborate this information. I directed the interpreter to state to them that while they conducted themselves properly they would be treated with kindness and consideration, but that if detected in horse stealing or other crimes against the whites they would be sure to meet with the severest punishment and that for the purpose of protecting their interests as well as its own, the government was about to establish a post and send soldiers to reside in their vicinity. They appeared to understand this perfectly but I am inclined to think it made but little impression upon them, and I was unable to obtain from any of them the name of their captain or the entire number of the tribe. After remaining about an hour at the Rancheria we returned to camp having asked two or three who spoke Spanish to accompany us, which invitation they declined. I suspect that nothing but our numbers and the well armed condition of the party prevented our being treated with incivility.

We left Lieut Moore's encampment early on the 8th taking with us but one pack mule and the cart which was necessary on account of the attached viameter and proceeding S.S.W. 34 miles encamped upon a small stream called Cottonwood creek, having crossed early in the day at a distance of eight miles from Moore's encampment another small stream which I called Gopher Creek. The latter I think is dry during the summer months but the former is undoubtedly constant during the year. Our route this day was over the most miserable country that I ever beheld. The soil was not only of the most wretched description, dry, powdery and decomposed but was everywhere burrowed by gophers and a small animal resembling a common house rat, of a whitish gray color, short round body and very strong bony head which I had never seen before.[27] These animals are innumerable though what they subsist upon I cannot conceive for there was little or no vegetation, and their holes and burrows into which a horse sinks to his knees at almost every step render the travelling difficult and dangerous. The low hills south

[27]Derby is likely referring to a "pocket Gopher," known scientifically as *Thomomys bottae*. It is common to Los Angeles County. See the L.A. County Department of Agriculture website.

of Gopher creek extend about 8 miles into the plain, their summits being about on the same level as the plain between the stream and Moore's creek, and are singularly intersected by valleys running nearly S. and N. which are crossed by other numerous small valleys running nearly E. and W. thus dividing the whole of this portion of the valley into blocks of hills about ¼ of a mile square and from 1 [100] to 300 feet in height. In riding through these valleys, the country presents the appearance of a large city which has been partially overwhelmed by the ashes of a volcanic eruption which resemblance is heightened by the continual cropping out of upright strata of clay from the loose soil of the hills, resembling exactly the ruined walls of adobe houses.

Upon the 9th we arrived after a march of 13 miles upon the N. bank of Kern River a very broad and deep stream with a current of six miles an hour which rising high up in the Sierra Nevada discharges itself by two mouths into Buenavista lake near its northern extremity.[28] Three large sloughs also make out from the river near its mouth and form an extensive swamp in the plain upon the north bank of the lake. We found the river impassable with animals and not time to make a sufficient raft. I therefore concluded to examine its Northern bank on the 10th and proceed to the Northern extremity of the Lake for the purpose of ascertaining its Latitude and Longitude as well as its general character and appearance. Buenavista Lake is a sheet of water about 10 miles in length and from 4 to 6 in width. It lies about 8 miles from the head of the Tulares valley formed by the junction of the ridges of the Coast range and Sierra Nevada, in latitude — N. and Long. — E. [Derby likely expected to add the numbers later] Like the other bodies of water in the valley it is nearly surrounded with Tule and upon its N. and E. banks there is found a heavy growth of willows. A slough some 60 miles in length connects it with the swamps and bodies of standing water in the bed of the Ton Taché and through them with the great Northern Lakes. The surrounding country is sterile and unproductive where not

[28]The Kern River, of some 165 miles, actually has a southern course, discharging water from the high Sierra Nevada. It turns west just east of Bakersfield, and originally discharged its water into Buena Vista Lake. The lake, like Taché Lake, has disappeared as a result of irrigation. The discharge from Kern River is used both for agriculture and urban use today. See David Carle, *Introduction to Water in California* (Berkeley: University of California Press, 2004), 40–42, 70–75.

an absolute swamp, with the exception of that portion lying immediately at the head of the valley and among the ridges of the surrounding hills, where are some extremely fertile and well timbered spots, two of which are now occupied and cultivated as Ranchos or farms. Nothing can be conceived more inappropriate than its name, for no place can be imagined more forlorn or desolate in aspect. Even at the early season of our visit, clouds of the most venomous mosquitoes tormented us during the day and goaded us to madness during the nights and we found here scorpions, centipedes and a small but extremely poisonous rattlesnake about 18 inches in length, reptiles which we had not before noticed in the valley and which with the gophers and ground rats were the only denizens of this unpleasant and uninhabitable spot.

A road from Los Angeles passing through what is called the Tejon at the head of Valley crosses Kern River 13 miles above its junction with the lake and continues through the valley to the eastward of the lakes as far North as the mining district upon the Stanislaus River.[29] On the 11th we started on our return to Moore's Creek taking a path about 15 miles W of that we had followed in leaving, for the purpose of more thoroughly examining the country. We found it however of precisely the same character throughout, barren decomposed soil, and no trace of vegetation but a few straggling Artemisias, except upon the margin of the creeks. Gopher Creek we found entirely dry at the lower point of crossing, its shallow stream sinking in the sandy soil some eight or ten miles above. We reached Lieut Moore's encampment about 4 P.M. on the 12th somewhat fatigued and excessively hungry after our disagreeable march, for in the hurry of leaving our pork and coffee had been forgotten and as we had seen no game upon the road we were obliged for four days to subsist entirely upon the hard bread, with which, reduced to a fine powder by its jolting on the mule, we made with water a species of poultice which had served but as an indifferent palliative to the cravings of our appetites.

We immediately set about making preparations for our march to the north and having no reason to apprehend danger from the Indians I

---

[29]Tejon Pass was well known at the time, being some fifty miles north of Los Angeles on what is today Interstate 5.

concluded to take but six of Lieut Moore's command with me as an escort, & those but to King's River intending to dispense with the services of the escort entirely from that point. But on arriving at Tule River which at the point of crossing we found on the 13th to be but 7.16 miles from Moore's creek I met a Mr. Shumway who with a single Indian was travelling South to establish a ferry upon Kern River, and he informed me that there was not the slightest danger to be apprehended upon the route, the Indians being all friendly, he with a single companion having travelled through them from the Mariposa river, I deemed it unnecessary to require the services of the escort beyond this point.[30] Accordingly, after sending back to the encampment for the wagon floats which were required in crossing, the river being here about 250 feet in width and extremely rapid and getting my party safely upon the Northern bank, Lieut Moore returned with his party to their encampment and took up their line of march for Monterey. The country had improved much in aspect from the time of leaving Moore's Creek and upon Tule River, and during our march upon the 14th we passed over rich tracts of arable land, fertile with every description of grass and covered in many places with a fine growth of heavy oak timber.

I observed that Tulé river might be easily bridged it being then at its highest stage and not subject to overflow, and upon its banks could be found plenty of timber available for the purpose, while the rocky hills some 5 miles above the crossing would furnish stone for the abutments and a pier which last could be erected upon a small island of hard gravel, covered at present with willows and situated at this point, near the middle of the stream. Upon the 14th we arrived at the river Frances a large stream 19 miles to the North of the Tulé. This stream flows nearly exactly from the hills and empties into the Taché Lake about 20 miles North of its S Eastern extremity.[31] It is, at the point

---

[30]Mr. Shumway cannot be identified. Mariposa River, or creek today, was the site of a large mining operation. The creek ran south, then east, into the San Joaquin River, just south of Merced, California.

[31]Derby's Frances River, like so many originating in the Sierra Nevada, has disappeared due to irrigation. It is identified as the "Four Creeks" by Carson, Derby's companion on his first survey trip. Farquhar identifies it as the Kaweah River. See Doris Shaw Castro, *James H. Carson's California, 1847–1853* (New York: Viking Press, 1997), 44; Farquhar, "The Topographical Reports of George H. Derby," 80.

upon which we arrived, divided into five branches or sloughs four of which separate from the main river about 4 miles above, joining again from 1 to 10 miles below, while the southern branch has a separate and distinct course from the Sierra and joins the main stream in the marshy ground near its junction with the lake.

The country some eight miles in length by six in width contained between these branches is a beautiful smooth level plain, covered with clover of different kinds and high grass and thickly shaded by one continuous group of very heavy oaks of a larger and finer description than any I have seen in the country. At this time each of the creeks was at its height, deep, rapid and four of them much wider than the Tulé river, their beds were also obstructed with floating and sunken timbers, previously felled for bridges which the high stage of the water rendered useless, and we should have found it difficult to cross had it not been for the kind assistance of the neighboring Indians who flocked to meet us and eagerly went to work carrying instruments and other property upon their heads and swimming the animals and the little cart across without the slightest accident occurring.[32] These poor people accompanied us from creek to creek assisting us cheerfully at each crossing and in return for their kindness I gave them a portion of our provisions and some dirty shirts, stockings and worn out boots which they prized as articles of exceeding luxury and accepted with vast satisfaction. There are two large Rancherias upon the river numbering together about 400, they are however apparently quite separate and distinct one occupying the southern the other the two northern branches.[33]

Nothing could exceed the kindness and hospitality with which they treated us, and I gave the Captains certificates to that effect which they will undoubtedly present to all future passers by. The name of the 1st Rancheria as nearly as I could write it from their pronunciation is the He-ame-e-tak's, their captain's Dec-e-jah, the 2nd Rancheria is the Cowees, their captain Francisco, an old mission Indian from San

[32] The lower portion of Tule River has disappeared due to irrigation. But two upper branches still flow, providing water for Lake Success, east of Porterville, California.

[33] These Tule River Indians were Yokuts of either the Tulamni or Tache tribes. See Wallace, "Southern Valley Yokuts," 448–61.

Luis Obispo.[34] They called the river the Ee-dek, but I think the name of Frances much more euphonious. There were many mission Indians among them, some from the old mission San Juan, some from San Lorenzo, but mostly from San Miguel and San Luis Obispo. All of these spoke Spanish a little. They appeared remarkably healthy and though by no means beautiful were comparatively well favored. The soil between the creeks is naturally, being so well watered and shaded, of the richest description as an evidence of which I may mention that I observed poles of willow stuck in the ground by the Indians as parts of rabbit traps, which had taken root and sprouted into trees. The distance of the 1st *Pyramid* creek (so called from the remarkable shape of a hill near which it rises) to the 2nd is .98 of a mile from the 2nd to the 3rd 1.04 miles, between the third and fourth (the last of which is the main stream) 2.81 miles, and between the 4th and 5th .85 of a mile. The third or main stream and the 4th are the widest and most difficult of crossing, the 2nd and 5th I believe to be dry during the summer but they may all be very easily bridged permanently, every material being ready upon the spot. I ascertained from the Indians inquiring that they were not subject to periodical sickness and they communicated to me much interesting and I believe reliable information regarding their numbers, manners and customs and those of the other tribes of which they had any knowledge.[35]

The 15th and 16th were occupied in crossing the creeks, which was a matter of labor and difficulty to accomplish, and in considering an examination of the country in their vicinity. On the 17th bidding adieu to our friends the Cowees, we started for King's River, our course lay over a barren sandy plain interspersed here and there with spots of vegetation but in general a perfect contrast to the rich soil upon the River Frances.[36] About six miles from the fifth creek we observed

[34]Derby is likely referring to the Dochiu, a Yokut tribe more often associated with Mill Creek, a tributary of King's River. Cowees is a corruption for Kaweahs, a Yokut tribe associated with Kaweah River, which drains into the lake of that name today. The lake is just east of Visalia, California. See Kroeber, *Handbook of California Indians,* 479–80; Wallace has the first group as the "Dachi" (see "Southern Valley Yokuts" 461).

[35]Derby likely used this "knowledge" to derive the estimates at the conclusion of the journal.

[36]King's River was named originally by Lieutenant Gabriel Moraga in 1805 as Río de los Santos Reyes. Its three branches break out of the Sierra Nevada east of Fresno, one branch eventually reaching the San Joaquin River. Other southern branches have disappeared due to irrigation.

four isolated hills or buttes about 800 feet high, two of them ranging nearly N & S. and about two miles apart, the other two about three miles farther to the East. At the first of these there is a constant spring of water and on the Eastern side of the hill some 300 yds from the spring a deep circular excavation or shaft with the remains of an old windlass lying in its vicinity. We were somewhat puzzled to account for the trace of human labor at such a distance from any settlement, but I presume it to be the remains of a well dug many years ago by the Spanish soldiers of the missions, who we know were accustomed to make excursions into these plains for the laudable purpose of seizing upon and Christianizing their wild inhabitants. We arrived upon the banks of King's river about five P.M. having travelled 23.66 miles in a direct line across the plain. We found here two enterprising men named Jones and Rider who with much difficulty had got a whaleboat up to this point and stretching a strong rope across the river had established a ferry. They informed us that there was another ferry about 5 miles above them kept by a man named Hampton who with another man and themselves formed the entire population of that part of the country.[37] We encamped at their ferry and crossing the party upon the 18th I took two or three men and rode up the northern bank about 15 miles to examine the river and the character of the country in its vicinity. The King's River is the largest stream in the valley, at this time about 300 yards wide with a rapid current and the water cold as ice. It is about 60 miles in length rising in two branches high up in the Sierra, which uniting about forty miles from its mouth flow in a S.W. direction through the hills and valley and empty into the Tache Lake at its N.E. extremity. Its banks are high and well timbered and the country in its immediate vicinity is apparently fertile. It forms five sloughs like the Frances, but they are much wider and the country between them is swampy and difficult of access.

---

[37]Jones's Ferry was still in service in the 1850s, located southeast of Fresno, California. It was used to cross one of the branches of King's River, which has long since disappeared due to irrigation. A Colonel Hampton had a second ferry over yet another branch of the river, five miles north of Jones's Ferry. They were apparently located between the two towns of Sanger and Reedley. See Mildred Brook Hoover, Hero E. Rensch, and William N. Abeloe, *Historic Spots of California* (Palo Alto: Stanford University Press, 1990), 84.

There are not less than 17 rancherias of Indians upon this river, numbering in all probably about 3,000 (including those situated among the hills in the vicinity). Of these, those living upon the lower part of the river are friendly and well disposed towards the whites, those high up among the hills are entirely ignorant treacherous and mischievous. I was informed by Col. Hampton at the upper ferry that the Cho-e-emince Rancheria situated in his vicinity and numbering some ninety warriors, had been quite troublesome of late, using his horses without permission, and in one instance attempting to take his boat to ferry over a party of Sonorans.[38] I thought his isolated position surrounded by these mischievous savages one of considerable danger but he appeared to feel no apprehension of their committing any overt act of hostility. On my return to the encampment I found a party of Indians armed with bows and arrows had passed down upon the southern bank, stopping at the ferryman's hut and telling the occupants that they must leave that part of the country in four days. The ferrymen were not at all alarmed however and I think that our presence and the Knowledge that a body of troops was soon to be stationed in their neighborhood had a very salutary restraining effect upon these Indians who I believe would become exceedingly troublesome if they had no dread of the consequences. It was this band that murdered Gardner and his companions in the Spring of 1849.[39]

We left the ferry on the 19th and travelling S.W. for 19.84 miles encamped on the edge of a swamp at a point about three miles above the mouth of King's River and immediately opposite the N [north] Tonto village.[40] As this point had been indicated in my instructions for examination as a suitable site for a military post, I was anxious to

[38]Reference is here made to the Choinimni Yokuts, the main tribe on King's River. Their village was well up the river where Mill Creek enters it, as Colonel Hampton suggests. The village was called Tishechu. See Kroeber, *Handbook of the Indians of California*, 474–75.

[39]William R. Gardner left Monterey on March 1, 1849, with six wagons loaded with goods. He reached King's River and passed up it a considerable distance, meeting ever-increasing numbers of Yokut Indians. They coaxed him into their village by showing him gold, then murdered him and four Sonorans who were with him. James H. Carson reported the murders in his *Recollections of the California Mines*, which has been reprinted as Carson, *Bright Gem of the Western Seas: California, 1846–1852*, ed. Peter Browning (Lafayette, Calif.: Great West Books, 1991), 16.

[40]The "Tonto" village is likely the Tanaiu village, Yokuts that lived below the Choinimni, near the mouth of the river.

cross the river and visit it, but was informed by the Indians a large body of whom swam across to our encampment that all the country in the vicinity was overflown [*sic*], and that it would be impossible to cross even if we were to construct "balsas" of tule owing to the rapidity of the current. It was evident enough that the country was overflown and as I found it impossible for anything but an Indian to get even to the bank of the river I was reluctantly obliged to give up my idea of crossing at this point. I could see enough of the country however, upon the other side to satisfy myself that it would be a very unpleasant place for anything but Indians to occupy and the fact of our not being able to get at it at all would be a sufficient proof of its non eligibility in a military point of view. The N [north] Tontos are the finest looking Indians I have met with in the valley, well built athletic men nearly six feet high, and their physiognomy struck me as less repulsive than that of the Tachés and other Indians whom we had seen. Their Captain was an old Indian named Antonio, he had a certificate of recommendation and offers of assistance to the extent of his abilities, appeared to be well merited. He stated the number of the N. Tontos at about 300, and informed us that he had never seen the waters so high as they had risen this season. I ascertained from him that the months of August and September are a very sickly period at this point, and was afterwards satisfied of this from information that I received from a respectable man who had crossed the King's river in the month of August at the village, and who found all its inhabitants prostrated by sickness to such an extent that he could get no assistance from them in crossing.[41] On the 30th we marched 24.92 miles on a devious course with the object of striking the "Sanjon de San Jose" which I presumed to be a slough connecting the River San Joaquin with the Taché lake but were unable to get further on account of the river, the ground between the lake and the San Joaquin being entirely cut up by small sloughs which had overflown in every direction making the country a perfect swamp which I found it a matter of great difficulty to cross.

We saw numerous bands of wild horses, numbering I should think in all more than a thousand they were at some distance but their

[41]The sickness here described was obviously malaria.

appearance rendered me extremely anxious for the safety of my animals, which would infallibly have been lost if they could have broken loose and joined them as they appeared much inclined to do. We were engaged on the 21st, 22d & 23d in getting through the river, crossing no less than eight distinct and pretty large sloughs one of which we were obliged to raft over, before arriving at the Sanjon.[42] In all of these sloughs a strong current was running S.W.ly or from the San Joaquin river to the lake. The country over which we passed between these sloughs was miserable in the extreme and our animals suffered terribly for want of grass, there being no wood upon the plain except an occasional willow on the largest sloughs we could make no fires and were consequently obliged to return to bread and water, a diet which though simple in the extreme I somewhat preferred to the raw salt pork on which the men luxuriated. The Sanjon de San Jose is a large and deep slough about 40 miles in length connecting the waters of the Taché Lake with the San Joaquin River with which it unites at its great Southern bend. At this time it was about 240 feet in width and with an extremely slow current setting toward the river. I do not think it possible to communicate directly with the lake through this slough, an attempt had been made a week or two previous to our arrival by a party of men in a whale boat who examined it for twenty or thirty miles and found it branching off into innumerable smaller sloughs which intersected the tule swamp in every direction. They also reported that there was a fall of water some six or eight feet at this point. I think it highly probable there may be a rapid near the mouth of the slough, but should find it difficult to account for a direct fall of that height, and as I could hear of no one else who had seen it although I met with many who had crossed the Sanjon in every direction during the dry season, I am inclined to disbelieve in its existence. The whole country some 40 miles in extent in a S.ly direction, by 10 in width, between the San Joaquin River and the Taché is, during the rainy season and the succeeding months until the middle of July a vast swamp, everywhere

[42]For the "Sanjon," see note 4. The Sanjon region, or lands around it, was given out in a series of Mexican land grants, starting in 1840, including that of Rancho Rincon de Sanjon, Rancho Sanjon de Santa Rita, and Rancho Sanjon de los Moquelumnes. There is little evidence of a serious attempt to occupy and develop these ranches.

intersected by sloughs which are deep, miry and dangerous. A wagon road therefore crossing the Sanjon at any point is impracticable, as it could not by any possibility be travelled more than three months in the year. I regret that we were unable to follow up the Sanjon to the lake, but this was utterly impossible. We found that we could not proceed a quarter of a mile in either direction without getting hopelessly mired, and as the water was fast rising we could not get back over the swamp which we had already with great exertion and difficulty crossed. There was neither wood nor grass where we struck the slough and though without any materials for building a raft it became necessary to cross it immediately. In this emergency a man of my party volunteered to swim the Sanjon and proceed on foot to the nearest ferry upon the River San Joaquin for the purpose of procuring a boat. This course was accordingly adopted and the boat, brought up by a Mr. Bridger from the upper ferry at the southern bend of the river arriving that evening, we succeeded in crossing the animals and arriving at the high land upon the bend early upon the 24th.[43]

The River San Joaquin rising in the spurs of the Sierra Nevada flows in a S.W.ly direction for some 60 miles to a point where making a great bend of over 90° it changes its course to the N.W. emptying finally into Suzuin [Suisun] Bay. At the time of our arrival its waters were at their highest stage, the back water from the river had filled the innumerable sloughs extending from its banks in every direction across the valley, and I was informed by parties arriving at the ferry in boats on their way to the mines that the whole country to the north was overflown and that large numbers of men and animals were encamping upon the plain finding it impossible to proceed farther by land, and being unable to cross the river. The point upon which we were encamped at the Southern bend was some ten feet above the level of the stream and for two miles in extent N. & S. was not overflown, a road from the debouche of the passes of Santa Anna and Pacheco distant about 25 miles might therefore, crossing here, be made available as a means of communication with the Eastern side of the Lake and the streams

[43]While the famous mountain man Jim Bridger spent time in California, the man mentioned by Derby was someone else, not identifiable.

flowing into it at any season of the year.[44] At this time, the river at this point was about 200 yards wide and from ten to twenty feet in depth but during the dry season it falls to from 4 to 2½ feet. I think however that a large boat could ascend the river *to the bend* during nine months in the year, its farther progress would be prevented at any time by the sand bar formed by the confluence of the Sanjon, which extends across the bed of the river and is covered but by about 2 feet of water at its highest stages.

Finding it impossible to proceed farther down the valley with advantage I started upon the 27th from the San Joaquin to the South, skirting the western coast of the Taché Lake on my march. I had previously sent two men to Monterey with orders to proceed through the pass of Santa Anna, of which no examination has yet been made, and to ascertain whether or not a wagon road might be made through it. At a distance of 22 miles from the San Joaquin we crossed the dry bed of a stream flowing during the rainy season from the coast range to the lake, upon the bank of which we found a few cottonwood trees and a little grass but with this exception the whole valley south of the bend, between the Coast Range and the Taché lake we found a miserable barren, sandy desert, with no vegetation but a few straggling artemisias, and no inhabitants but a few attenuated rabbits and gophers. We encamped upon a slough connecting with the Sanjon upon the 28th where I observed some singular ducks, of a bright red or claret color with a white bill and short black tail and others of brilliant black plumage with red bill and feet, resembling very much turkey buzzards in their general appearance. I was anxious to obtain some specimens of these birds but although we shot several we were unable to get them from the water. We encamped upon the shore of the lake upon the 29th having made 17.32 miles and on the 30th reached the debouche of the pass of San Miguel having travelled 24.60 miles passing on our route the Rancheria of the Tente Taché's situated upon the N.E. extremity of the lake and containing some 250 or 300 inhabitants.[45] A few of them visited our encampment,

[44]These two passes were just east of Mission San Juan Bautista. Pacheco Pass was named for Francisco Pacheco. See Albert Shumate, *Francisco Pacheco of Pacheco Pass* (Stockton, Calif.: Holt-Atherton Pacifica Center for Western Studies, no. 7, 1977).

[45]See notes 21 and 22. Derby gives this band a larger population this time.

two of whom could speak Spanish and left well pleased with the treatment they received. We took up our line of march for Monterey by the same route that we had pursued in coming out, upon the 31st and arrived there without incident upon Saturday the 8th of June.

The Tulares valley from the mouth of the Mariposa to the Tejon Pass at its head is about 120 miles in extent, and varies from 8 to 100 miles in width. With the exception of a strip of fertile land upon the rivers emptying into the lakes from the East it is little better than a desert. The soil is generally dry, decomposed and incapable of cultivation and the vegetation consisting of artemisias and wild sage is extremely sparse. The only point in the whole valley which struck me as at all suitable for a military post was the small portion of interval land contained by the five creeks of the River Frances.

A position here would be central, being easy of communication with the King's River to the north and with Kern river to the south upon which two streams and their tributaries are situated the greater number of Indian Rancherias in the valley. The land is excellent for cultivation, well timbered and abundance of building material may be found in the vicinity either stone or heavy pine and oak timber. A road leading through the Tejon pass from Los Angeles, and intersecting the emigrant trail through Walker's Pass near Kern river passes directly through this point to the northern mines of the San Joaquin valley.[46] This road will undoubtedly be much travelled when brought into notice, and the post being established at this point will contribute much to its safety and protection. The post could be supplied by wagons through the pass of San Miguel either from San Luis Obispo or Monterey, or by a wagon road from Monterey through the pass of Pacheco, or Santa Anna, crossing the San Joaquin at its southern bend and the King's river at the lower ferry. From the information regarding the character of the San Joaquin between Stockton and the Southern bend I have no doubt that it may be navigated by small steamboats from the latter point during eight or nine months in the year. In this case, this

[46]Walker's Pass was shown to Joseph R. Walker by Indians in 1843. It became a major emigrant route, passing over the Sierra Nevada just northeast of present-day Bakersfield. Its route is roughly that of California Highway 178.

would be by all means the most desirable route of communication as a depot might be established upon the high land at the southern bend which would be but two short days journey from the post.

I should judge that the Lake from the difficulty of getting to its shores, and the swamp and sloughs that every where environ it could never be made available as a means of communication, though if this were possible, a ferry upon its waters would shorten the distance to its East bank by over fifty miles.

The whole number of Indians in the valley as near as I can judge is about 4,000.[47] These are by no means connected with each other, the Rancherias upon the southern rivers being in fact ignorant of the existence of those situated farther to the north. Those who have had an opportunity of becoming acquainted with the whites are friendly and well disposed, the others although ignorant, mischievous and perhaps malevolently inclined may easily be kept in subject by a small force, and by kind treatment. [I] have the honor to enclose a map and [sketch] of the valley showing the various routes and the geographical position of the points referred to in this report also two tables, one of altitude and longitudes the other of distances as measured by the viameter, all of which I hope may prove satisfactory to the General Commanding.

I am sir
With high respect
Your obdt servt
Geo H. Derby
Bvt 1st Lieut Top Engs

Maj. E. R. S. Canby
Adj Genl 10th mil. Dept.

[47]Derby is here referring to the region south of the American River, including especially the San Joaquin River valley and the Tulares region. This population figure was later confirmed by G.W. Barbour, who made treaties with these Indians in 1851, as well as by Captain E.D. Keyes, who led Barbour's escort. See Barbour to Commissioner of Indian Affairs, May 14, 1851, NARG 75, LR, California Superintendency; Keyes to Assistant Adjutant General June 17, 1851, NARG 393, LR, 10th Military Department. Such evidence, coming as it does from government officials charged with controlling these Indians, offers a much more accurate depiction than the newspaper accounts and select primary documents consulted by Kroeber, Heizer, and Cook.

CHAPTER 4

The Journals of Williamson, Lyon, Wessells, and Keyes 1850–1851

The following series of journals and reports were written in 1850 and 1851, just after Ord's initial expeditions into the regions around Los Angeles and Derby's three surveys of the Central Valley. They were conducted by topographical engineers, such as Lieutenant R. S. Williamson, as well as regular army officers, including Captains Nathaniel Lyon, Henry Walton Wessells, and E. D. Keyes. While Williamson had training as a topographical engineer, the other three officers did not. All four of these men kept extensive journals, similar to Derby's. Unfortunately, maps of their expeditions have survived in the army records in only two cases, that of Lyon and Williamson. They are reproduced herewith.

All four of these officers had standing orders very similar to Derby's. The army commanders in California wanted to know where the best locations were for military forts, how difficult it would be to supply them, what kind of building materials existed at the various locations, and the health of the climate at the proposed sites. In particular, the officers also received orders to determine the extent and nature of the local Native populations. Lyon actually left three journals, two of his expedition into the Clear Lake region and a third of his attempt to punish the Indians of northeast California who had killed Lieutenant Warner. Both Wessells and Keyes accompanied government commissioners who were negotiating treaties. When all these journals—six in number—are placed in conjunction with the wealth of information collected by Derby and Ord, a very good picture of early California emerges, complete with a careful rendering of the populations of Indians who were in the mining districts as well as those who were in isolated, and unexplored, parts of the state. As with Derby's and Ord's reports, these journals demonstrate that far fewer Indians were found in much of the relatively unexplored regions of California than previously thought.

Williamson's account is an extension of the Warner Expedition account into northeastern California, the first such survey of the region by government officials. John C. Frémont had visited the northern portion of the Central Valley several times in former years, but the map that resulted from his trip had little detail on the northeastern mountains, the Sierra Nevada of California, and Frémont provided little information relative to the Native occupants. Williamson successfully collected the maps and notes made by Captain William H. Warner after Indians ambushed and killed him. Williamson was interested in determining both the locations and numbers of American Indians in the mountains surrounding Goose Lake, and he also was charged with producing a map of the region, which is included here. Warner, and later Williamson, had received orders to survey the region for a possible railroad route through the mountains.

The two journals of Lyon's campaign into the Clear Lake region north of Sonoma, called the "Clear Lake Expedition," while providing considerable information about the region and the Indians, also resulted in two terrible massacres of Indians, only one of which is known to history today. It is called the "Massacre at Bloody Island." Lyon's first report of the engagement—roughly a third the size of his final report—was sent off to Washington, and a brief, edited version of just two or three pages was published in the serial set and later published by Robert Heizer. A complete version of the first report is included here. The second journal, much longer and more detailed, offers the only complete description of the assault by Lyon's troops on the island, then proceeds to give a detailed description of how his forces followed the Pomo Indians into the Russian River valley and attacked more villages, killing Natives and burning towns as he went. It was the single most destructive campaign conducted by the army against Indians in California history, but it has received very little notice in general histories of the state.[1] Lyon's two reports of the engagement are published here for the first time. Also included is Lyon's report of an attempt to punish the Indians of northeastern California, called the "Pit River Expedition." What is significant is the fact that Lyon found virtually no Indians at all near California's northern border with Oregon.

The journals of Wessells and Keyes are significant for a number of reasons. They entered regions where there had been some conflict between ranchers/miners and Native Americans, and they accompanied commissioners who were charged with negotiating treaties with those same Indians. Keyes spent part of the summer of 1851 with George Barbour and Oliver Wozencraft, who signed treaties with virtually all the bands in the southern mining district, namely, the upper San Joaquin River basin and well to the south of it. Keyes revisited the region that Derby went through and reconfirms Derby's assessment of Native populations, offering a figure close to the 4,000 Indians that Derby reported.

Wessells accompanied Redick McKee into the Russian and Eel river valleys, then northward into the Klamath, Trinity, Scott, and

[1]The two most often cited surveys of California fail to even mention Lyon's engagements with the Pomo Indians. See Kevin Starr, *California: A History* (New York: Modern Library, 2005), and Andrew Rolle, *California: A History* (Wheeling, Ill.: Harlan Davidson, 2003). Indian historian George H. Phillips uncovered a brief newspaper account of Lyon's assault on the island. See Phillips, *The Enduring Struggle: Indians in California History* (Sacramento: Boyd & Fraser, 1981), 44–45. Robert Heizer left a typescript manuscript entitled *Collected Documents on the Causes and Events in the Bloody Island Massacre of 1850* (Berkeley: University of California Archaeological Research Facility, 1973). He discovered Lyon's brief report of the expedition, which had been edited down to two pages, which were reproduced in Senate Executive Document 47, 31st Congress, no. 1, part 2, 81–83. He includes other documents, such as the same newspaper account discovered by Phillips and found in the *Alta California*, May 28, 1850.

Salmon river valleys. The party found very few Indians in the Russian River valley and only a few hundred in Round valley. Vague reports had suggested much larger populations in the Klamath, Trinity, Scott, and Salmon Rivers, but Wessells was convinced that the entire Native community living in these valleys—or the Native population of all lands north of Round Valley—was just five thousand.[2]

Just before his death, Warner supposedly found a usable railroad route far north of Goose Lake in southern Oregon. Williamson reported the route but apparently never saw the pass that was located on the map produced from the expedition. It mattered little, as nothing ever came of it. Williamson later conducted several major railroad surveys to examine the various western routes considered by the War Department in 1853, most of which were far better suited for roads than the route found by Warner.

General Persifor Smith was quite pleased with Warner's discovery and assumed that it would offer a gradual and less demanding route for a railroad into northern California. His report of the expeditions—both Warner's and Williamson's—follows.

General Smith's Preface to the Williamson Report

Headquarters Third Division[3]
Benicia, February 27, 1850

Colonel:

I have the honor to transmit the report of Brevet 2nd Lieutenant R. S. Williamson, of the topographical engineers, of the reconnaissance made by the late Brevet Captain W. H. Warner of a route through the Sierra Nevada by the Upper Sacramento.[4] It will be seen that from the waters of the bay of San Francisco, following the valley of the Sacramento on the west side, to the foot of the hills at the base of the Sierra, is a smooth inclined plain, without obstacle. Following up

---

[2]Other reports were filed by the Indian commissioners who accompanied Wessells and Keyes. They can be found in NARG 75, LR, California Superintendency.

[3]General Smith's letter forwarding Williamson's report is found at NARG 393, LR, DP.

[4]Lieutenant Williamson entered the Topographical Engineers after graduating from West Point in 1843, fifth in his class.

Pit River, or Cow Creek, and then crossing to Pit River a route can be carried through to the eastern foot of the Sierra, with a gradient of thirty-eight feet per mile—thus passing from the Pacific to the eastern base of the Sierra Nevada almost without any of the obstacles common to most of the roads in the United States.[5]

Captain Warner's death cannot be too deeply lamented. He was devoted to his profession, and well qualified, by his perseverance, habits of endurance, and his uncommon accuracy, to reach an eminent position in it.[6] He and Lieutenant Williamson were subjected to great hardship and suffering from sickness and want of wholesome food, and their labors were performed under great disadvantages. Lieutenant Williamson deserves great credit for the manner in which he has concluded the work and put on record the valuable result of their reconnaissance.

With respect, your obedient servant,
Persifor F. Smith
Bt. Major Gen commanding Division.

R. S. Williamson Explores the Upper Sacramento to the Pit River August–October 1849

Bt. Lieut. Col. Wm. G. Freeman[7]
Ass't Adj. Gen., headquarters of the army.
Sonoma, California, February 14, 1850

Sir:

Orders from the headquarters of the Pacific Division, dated June 27, 1849, were issued to Captain W. H. Warner, of the topographical

[5]Cow Creek is a tributary of the Sacramento River that runs just northeast of present-day Redding, California. A fort was constructed on the creek in May 1852, called Fort Reading, in honor of an early rancher in the region, Pierson B. Reading. Reading's name was also used to identify the steamboat landing on the river. When the railroad came through in the 1870s, the depot and later the town of Redding were established, named after a railroad man, Benjamin B. Redding. From the fort's headwaters, the expedition moved a few miles north into the Pit River valley, which had its source in Goose Lake, well to the northeast.

[6]Captain William Warner's death is described in detail in Williamson's survey report that follows.

[7]Lieutenant Williamson's original report of February 14, 1850, is found at NARG 393, LR, DP. An edited version was published as Senate Executive Document 47, 31st Congress, 1st session.

engineers, to take charge of an exploration from the upper Sacramento, across the Sierra Nevada, to the Humboldt River; the main object of the expedition "being to discover a railroad route through that section of the country." At the very moment it had been fully demonstrated by Captain Warner that such a route across the Sierra Nevada was practicable, and when nearly all the required information had been obtained by him, he was murdered by Indians. In consequence of this lamentable occurrence, the command of the topographical party devolved upon me, and it now becomes my duty to submit to you, for the information of the major general commanding the division, the following report, which imbodies [*sic*] as much of the information collected by Captain Norris as came to my knowledge.[8]

Four commissioned officers and eight men were ordered to compose the escort of Captain Warner, and the party was to move from Benicia on the first of July, with supplies for four months. The detail for the escort was not received by the officers ordered till some days after the first of July, and the collection of the different detachments composing it, with their preparation for the transportation of the supplies, &c., consumed so much time, that the escort did not start from Benicia till about the first of August, arriving at Sacramento City on the third. The escort was commanded by Brevet Colonel Casey, of the 2nd Infantry, who was accompanied by Lieutenants Shureman and Gardner, of the same regiment, and Dr. Hewitt, of the medical staff.[9] Captain Warner and myself, with three men, left Benicia on the 28th of July, arriving in Sacramento City and Sutterville on the 31st. Here Captain Warner hired a party of ten men, and also engaged the services of François Bercier, (ordinarily called Battitu,) as guide.

[8]Captain John S. Norris came to California as part of the New York regiment that arrived in 1847. He stayed on, later served in the Union army, and was captured during the Civil War in South Carolina.

[9]Colonel Silas Casey was a career officer who later commanded Fort Steilacoom, near Puget Sound, Washington, and received the brevet rank of major general during the Civil War. Lieutenants James Wall Shureman and W. N. Gardner came with the regiment to California from New York. Shureman died of blood poisoning in San Francisco in 1852. Henry Stuart Hewitt, a contract surgeon, was with Warner when he was killed. He later served as a surgeon in the Civil War, developing a close friendship with General William Tecumseh Sherman. Hewitt no doubt met Sherman when the young lieutenant was post adjutant at Monterey in 1849.

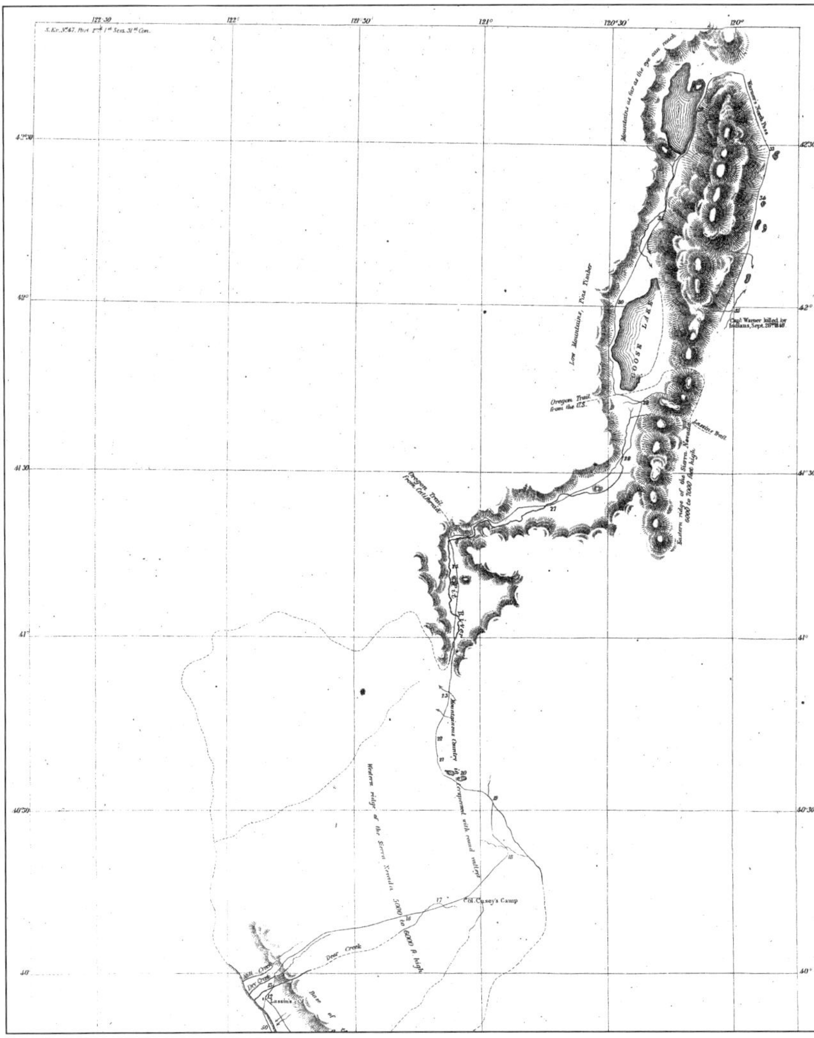

"Sketch of the Route of Capt. Warner's Exploring Party in the Sacramento Valley and Sierra Nevada, during the months of August, September, and October, 1849." From R. S. Williamson's journal of the survey of the Pit River and Goose Lake region, made to find a railroad route through the mountains, August–October 1849. *Courtesy of National Archives and Records Administration.*

This man was a very intelligent half-breed, born on the Red river of the north; had served some time with the Hudson's Bay Company in Oregon; had trapped in the rivers of the country we were about to explore, and proved himself to be a very valuable man for the short time his services were given to us.[10]

On the 13th of August we commenced our march up the Sacramento valley, on the east side of the river. Our party consisted of eleven hired citizens and two soldiers, and we carried with us a light wagon, a cart, and a few pack-mules. As there was no need of the escort until we should be among the hostile Indians of the mountains, we did not delay our march in order to keep company with it, but advanced to Lassen's rancho, on Deer Creek, where we remained a few days to jerk beef.[11] Captain Warner here gained a great deal of information with regard to the route across the mountains, from Mr. Lassen, who had guided a party across the country from Missouri in the summer and fall of 1848; had deviated, near the great bend of the Humboldt river, from the previously travelled road, and after much difficulty succeeded, with great good fortune, in finding a continuous ridge from the summit of the Sierra to the Sacramento valley just at the place where his own rancho was situated. He represented this to be an excellent road after the first ten miles, and expressed his opinion that it was a practicable railroad route. This induced Captain Warner to strike into the mountains from this point, instead of proceeding further up the Sacramento valley. The escort arrived at Lassen's three days after our party, and it was found that it was impossible to proceed into the mountains with the whole command, and move with sufficient rapidity to return

[10] François Bercier was born in 1821 to Pierre Bercier who worked for the Northwest Fur Company at the time. The Northwest and Hudson's Bay Companies were merged in the year of François's birth. Pierre may have been working along Red River when François was born, but Pierre spent most of his time in what was called the Columbia District, or northwestern Montana and Idaho, where François learned his trade.

[11] The rancho of Peter Lassen was a common stopping point along the upper Sacramento River. Lassen became a Mexican citizen in 1840 and received a grant of 22,000 acres four years later. He later pioneered the Lassen Trail from Missouri, which ran just south of Goose Lake, into the Sacramento River. Lassen died in 1859, and the ranch was ultimately purchased by Leland Stanford, who deeded it to Stanford University. See Mildred Brooke Hoover, Hero Eugene Rensch, and Ethel Grace Rensch, *Historic Spots in California*, ed. William N. Abeloe, 3rd ed. (Stanford, Calif.: Stanford University Press, 1966), 548.

before the snows commenced. Captain Warner, therefore, made a requisition upon Colonel Casey for a detachment of a commissioned officer and twelve men, which, with the citizens he had hired, would make his party sufficiently strong, as he thought, to resist any attack that might be made upon him by the Indians. Animals were required from Colonel Casey to pack our provisions and mount the infantry detachment, and this left him with very small means of transportation. Captain Warner had, by this time, given up all idea of being able to reach Humboldt River, and he proposed to follow the Lassen trail till he struck the Pit River, then, follow up this river to its source, and endeavor to find a pass through the eastern ridge of the Sierra near that point. Colonel Casey, in the mean time, was to advance slowly with the provisions, and was expected to be on Pit River to meet us on our return, and furnish us with supplies to enable us to explore a route either down that river or Cow Creek.[12]

On the 24th of August we left Deer creek and entered the mountains, party being about thirty-four strong, and having with us the wagon, cart, and about eighty-six animals. We found the hills quite steep, and the road very rough and rocky, so that it was with much difficulty we made nine miles, our mules being very poor, though the best that could be obtained. The next day, though the loads of the wagon and cart had been much reduced, we marched out nine miles farther, the wagon not arriving at camp till after dark, and the mules in the cart giving out entirely. We were exceedingly desirous of carrying the wheeled vehicles with us, as we wished to continue to measure the road with the odometer, but they retarded our march so much that Captain Warner determined to send them back, although it was practicable to take them along at a very slow rate. Our men were here taken sick hourly, and the next morning nearly half the party was unfit for duty.[13] Captain Warner had had a fever for several days, and was now quite sick, and in fact it was impossible for us to

[12]The route is apparently up Deer Creek and into the mountains, Deer Creek being nearly fifty miles south of Cow Creek. The party then turned north, examined Goose Lake, and returned via Pit River and Cow Creek, where Colonel Casey was to meet them with supplies.

[13]Malaria had been contracted while in the Sacramento valley. The entire party was soon down with "fever and ague."

advance immediately. During our second day in the mountains we met several men from the United States, and learned, to our surprise, that they were the advance of from ten to twenty thousand emigrants, who were coming into the country on this trail. The advance parties were composed mostly of men who had left their trains, under a false impression as to the distance to the valley, and they were in a starving condition.[14] They came to us in such numbers, begging for provisions, that our supply was materially diminished. As soon, however, as we met the trains, we found the emigrants generally pretty well provided. Captain Warner succeeded in hiring several of them to supply the places of those of our party too sick to advance; and leaving seven men in charge of the surgeons, and sending back our wagon and cart by one of the emigrants, with everything that we could dispense with, we left our camp on the fourth of September, and on the evening of the fifth we reached the headwaters of Deer creek, about fifty miles from Lassen's. Here we were obliged to wait another day, and to leave behind five more men, too sick to proceed. As the detachment of soldiers under Lieutenant Gardner was now reduced to ten men, and as he himself was unwell, he remained in camp with the sick. It is a singular fact that, during the first ten days in the mountains, not a single one of our party escaped an attack of fever.

Upon leaving the headwaters of Deer creek, we continued to follow the trail, crossing the headwaters of Feather river, striking Pit River on the 13th of September, and reaching the headwaters on the 18th, which are near Goose Lake, a body of water over twenty miles long, and impregnated with salt.[15] The impression produced by Mr. Lassen, as to the character of the road, was decidedly incorrect. For thirty-six miles after entering the mountains, the road was very rough, stony, and steep, continually ascending till we had attained an elevation of

[14]It was commonly the case that emigrant parties coming over the mountains found virtually no game whatsoever and arrived in a starving condition in California. On this occasion, though, the train that was behind the men still had supplies on their wagons, being better prepared for the trip than most wagon trains.

[15]The march overland is nearly due north, through the Lassen National Forest of today. It is worthy of note that over the one hundred or so miles of this trip through the mountains, the party saw no evidence whatsoever of Indian occupation.

over five thousand feet. The road followed a ridge between Mill Creek and Dry Creek, which streams flowed in steep ravines on either side, into which it was necessary to descend to get water. We seemed at the end of thirty-six miles to have arrived at the summit of a continuous range of mountains, from which we descended about two thousand feet in the next ten miles. It appeared certain that no railway could be constructed over the road we passed, unless extensive inclined planes were used. Between this mountain range and the range in which Pit River rises, the country is mountainous, but interspersed with numerous valleys from five to twenty-five miles in length, and between three thousand and four thousand feet above the Sacramento valley at Lassen's. The mountains near the source of Pit River are between six and seven thousand feet above Lassen's, and, as far as we could ascertain, formed a continuous unbroken range.[16] This range had been considered by Captain Warner as presenting the greatest obstacle in his way; and as it was impossible for any railroad to cross it where the Lassen trail crossed, he determined to pursue his examination to the northward, in hopes of finding a depression in the mountains. As many of our mules were worn out, he determined to take a small picked party, leaving the rest with me in camp; and accordingly he started on the 20th September with nine men, including the guide, and ten days provisions. He passed to the northward, leaving Goose Lake and the mountains on his right, and, after travelling seventy miles, passed through a deep depression in the mountains, which differed very little in altitude from the valley of Goose Lake.[17] The result from my calculation of the barometric observations gives the altitude of the highest point passed over by Captain Warner but one hundred and seventeen feet higher than over [our] camp at Goose Lake. It is true that a difference in the state of the atmosphere, at the time of observation, may have caused an erroneous result; but as no such difference

[16]The party apparently moved along the western edge of the Warner Mountains, Eagle Peak being the tallest at 9,892 feet.

[17]Warner had reached the vicinity of Lake Albert in Oregon, turning east and traveling through a depression; his small party then turned south, skirting the eastern shore of Goose Lake. Williamson's map of the region calls the "depression" that Warner traveled through "Warner's North Pass."

was apparent, the error could not be great. The discovery of this pass, then, makes a railroad route perfectly practicable, as far as the eastern or main spur of the Sierra Nevada is concerned.[18]

After going through the pass, there is no impediment until the road reaches the point where our trail first struck Pit River. This river, which is the main fork of the Sacramento, is said to have taken its name from the numerous pits on its banks, which are sometimes twenty feet deep, and which had been dug by the Indians as traps for deer, bear, &c. It is, as far as we followed it, a slow, sluggish stream, descending about eight feet in the mile, and having ample room between its banks and the hills for the construction of a good road. The banks in a few places show signs of being overflowed in the spring, but generally appeared high enough to confine the waters. The hills near the banks were not cut up by ravines, and we observed no dry beds of streams near the banks which would require to be bridged. The hills sometimes approach very near the bank and sometimes diverge, forming fine valleys. Where the trail to Lassen's leaves the river, it strikes into steep hills, where it would be very difficult to construct a railway. In the opinion of Captain Warner, the road should follow Pit River, or, by crossing over to the headwaters of Cow Creek, follow down that creek to the Sacramento valley. The altitude of the point where the road leaves the river is three thousand eight hundred feet above Lassen's; and as the distance to the valley must be at least a hundred miles by either of these routes, there would be an average descent of thirty-eight feet to the mile, which is perfectly practicable for a railroad route. Which of the two routes above mentioned is preferable, can only be determined by a minute examination. Our guide, Battitu, said he had travelled down Pit River to the valley, and after passing the cañon of twenty miles, the banks were open. From what I saw of the cañon, I think a road can be cut through it with much less labor than is expended upon many of our eastern roads. The whole country through which the road would pass till it reached the valley is well timbered with pine, fir, and cedar; an abundance of burr

[18]The commentary suggests that Cedar Pass, just south of Goose Lake, which was referred to as being on Lassen's Trail, was too steep for railroad construction, given the technology of that day. A description is found in Hoover, Rensch, and Rensch, *Historic Spots*, 549.

or millstone is found, as well as another silicious [siliceous] rock, well adapted for road covering, which would make a good building-material for common purposes.[19] Some specimens of these stones, taken from different locations on Pit River, I present for examination.

When Captain Warner had discovered the pass, and reached the eastern base of the range, he travelled to the southward, intending to recross the mountain on the Lassen trail.[20] On the 26th day of September he was riding in company with the guide, a short distance ahead of his little party. They had descended a little ravine and were ascending the rugged hill on the other side, when a party of about twenty-five Indians, who had been lying in ambush behind some large rocks near the summit, suddenly sprang up and shot a volley of arrows into the party. The greater number of the arrows took effect upon the Captain and guide, and both were mortally wounded. The Captain's mule turned with him, and plunged down the hill; and having been carried about two hundred yards, he fell from the animal dead. The guide dismounted and prepared to fire, but finding he could not aim his rifle, he succeeded in mounting and retiring down the hill. He died the next morning. The party was thrown into confusion and retreated at once. Two men, George Cave and Henry A. Barling, were badly wounded. Cave died before reaching the valley, while Barling reached Benicia, was placed in the United States hospital under charge of Assistant Surgeon Deyerle, and has now nearly recovered.[21] Captain Warner's body was visited several times, and his note-book, &c., brought to me. The Indians who made this attack are supposed to be of the same tribe, and have the same manners and customs, as those in the immediate vicinity of Tlamath [Klamath] lake.[22] They caused a great deal of trouble among the emigrants by stealing their cattle in the night; and they acted with a great deal of caution, never showing themselves during the day. They have no other arms than

[19]Millstone, or *burr,* is stone suitable for grinding wheat. Siliceous ooze is a soft rock often associated with sediment that covers the ocean floor.

[20]Warner was attacked by Indians just north or where the trail comes out of the mountains at Cedar Pass.

[21]Both men appear to be civilians, hired as laborers for the expedition.

[22]The suggestion is that they were Modoc Indians, but no exact identification was ever made.

bows and arrows, and generally go entirely naked. They seemed to have been emboldened by the presence of so small a party so far from the emigrants' trail, and presented themselves in considerable numbers in the vicinity of Captain Warner's camp for several days preceding the attack. It is difficult to make an estimate of their numbers, but they certainly can form a formidable body. I met on Pit River a party under charge of Mr. Peoples, who had been sent by Major Rucker, with a supply of provisions to the relief of the emigrants.[23] He told me that several of his men, while hunting among the hills for some lost animals, had discovered a body of the Indians, which they estimated at five hundred strong.

The information contained in Captain Warner's note-book was so full that it would have been superfluous for me to have visited the pass he discovered. The lateness of the season, and the grass on the road having been nearly consumed by the animals of the emigrants, rendered it impossible for me to advance to Humboldt River. My party was not sufficiently strong for me to attempt to punish the Indians; and our provisions were nearly exhausted. For these reasons I deemed it necessary to return immediately to Lassen's. I sought for conversation with the most intelligent of the emigrants to ascertain the character of the road between the mountains and Humboldt River, and they were unanimous in asserting that there existed no serious impediment to the construction of a railroad at any part of it. Upon the return of the remainder of Captain Warner's party with their sad intelligence, I succeeded in making arrangements with some emigrants for transportation of the two wounded men, in wagons, to Lassen's; and, this being done, prepared to return with the remainder of the party. I fortunately succeeded, a few days afterwards, in purchasing some provisions from an emigrant, which kept us scantily supplied till we reached the valley, which we did on the 14th of October. On the 10th and 11th we had the first snow of the season. I was exceedingly anxious to return by Cow Creek; but, having no guide, and being so short on provisions, I thought it necessary to return by the same route we came. Colonel

[23]Major Daniel Henry Rucker had joined the 1st Dragoons in 1837. He soon was attached to the Quartermaster's Corps and later served with distinction in the Civil War as a major general. The Mr. Peoples referred to was likely a civilian.

Casey, who, with the escort, had reached the headwaters of Deer creek, and had remained there for some time in camp, had proceeded me a few days to Lassen's, and left that place on his return to Benicia. I here sent back three men to carry provisions to the wounded, and to bring them to Benicia. I left, for their transportation, the wagon which Captain Warner had sent back; and, with the cart and pack train, left Lassen's on the 16th, crossed the Sacramento near the mouth of Deer creek, and returned to Benicia on the western bank of the river.

I have the honor to present, with this report, a copy of the meteorological and astronomical observations taken upon the expedition, and a sketch of our route. I will state the data from which this sketch was made, to show the degree of accuracy of its different parts. In going from Benicia to Sacramento City, from there to Lassen's, and thence back to Benicia, the distances were measured with an odometer, and the courses taken with a prismatic compass. At each turn of the road a new course was taken and the reading of the odometer noted, while at the same time a sketch of the country passed over was made, and the position of the houses, rivers, &c., determined. Hence I had complete compass notes, with distances actually measured, which I plotted. The positions of Sacramento City, Benicia, and Lassen's were then assumed on the map, by their latitudes and longitudes, and neither of the distances between the points so determined differed two miles from the corresponding distances as plotted; a degree of accuracy which is the result of great care and attention on the part of Captain Warner. From what I have above stated, I think the position of the different points on our route, while in the Sacramento valley, are laid down with accuracy. I have studiously avoided putting down anything that did not come immediately under our observation—confining myself entirely to a sketch of the route, except in the single case where I have taken from a map made by Captain Warner that part of the Sacramento river below Sacramento city, together with the bays of Suisun, San Pablo, and San Francisco. The numerous streams running from the Sierra Nevada to the Sacramento were all dry at the season we crossed them, excepting those with names; but in the winter season are formidable rivers, which would require to be bridged if the road was made on that side of the river.

There are, however, but two streams above Sacramento City, which come from the coast range. The part of the route from Lassen's to Pit River is laid down only approximately, as the road was so winding, and the forest so thick, that it was impossible to get our course with accuracy. After reaching Pit River, however, we had the nearly open bank to travel upon; and our courses were again observed with care, and the distance estimated by noticing the time taken to pass over any course. The latitude of our camp 29, near Goose Lake, was ascertained from Captain Warner's observations, while its longitude was assumed approximately, by reference to Frémont's map.[24]

As the Lassen trail is now an emigrant trail from the United States, a list of distances from camp to camp, as far as we proceeded on it, may be useful, and I therefore submit it with the map.

I have the honor to be, very respectfully,
your most obedient servant,
R. S. Williamson
Lieutenant United States Topographical
Engineers.

Bvt. Lieut. Colonel J. Hooker,
Assistant Adjutant General, Pacific Division

Nathaniel Lyon's Expeditions against the Indians May–July 1850

After the murder of Lieutenant Warner, military officials in California concluded that certain tribes needed "chastisement." These included the Pomo Indians near Clear Lake, who had recently killed two ranchers, as well as the Indians responsible for the murder of Warner. This

[24]The reference is to Charles Preuss's famous *Map of Oregon Territory*, published in 1848. Preuss, a native of Germany, joined Frémont in 1838 and accompanied him on several of his expeditions into the West. Somber and humorless, Preuss committed suicide in 1854, but his maps gave Americans a whole new appreciation of the American West.

duty was assigned to Major Washington Seawell, who, in turn, gave the assignment to Captain Nathaniel Lyon. Lyon left three reports of his expeditions against the Indians, two relative to his attack at Clear Lake (often referred to as the "Massacre at Bloody Island"), and one an account of his expedition into the Pit River valley, where he encountered few Indians. All three are included below.

Given the serious problems with disease, especially malaria, most tribes in California were decimated by the time of Lyon's expeditions. This included the Pomos of Clear Lake, who numbered far less than a thousand people by 1850. While Lyon killed upwards of two hundred Indians in his various engagements with the Pomo, thereafter, most military officers concluded that the Indians were seldom at fault for the clashes that occurred between ranchers/miners and Indians. Many officers showed increasing sympathy for Native Americans, and Lyon's slaughter of Indians—no other word can describe it better—was the last engagement of its kind conducted by the army in the 1850s.

The orders organizing the Lyon Expedition were issued in March by Assistant Adjutant General Canby, in a letter to Major Washington Seawell, who then assigned Lyon to lead the expedition.[25]

General Canby's Orders to Major Seawell

Headquarters 10th Mil. Dept.[26]
Monterey, Cal. Mar. 9, 1850

Sir:

You have been assigned by the Brigadier General Commanding the Department to the command of the expedition about to be undertaken for chastising the Indians in the vicinity of Clear Lake recently engaged in the commission of murders upon the inhabitants of the district south of Sonoma [north of Sonoma] and also upon the

---

[25]Major Washington Seawell graduated from West Point in 1825, serving in the Seminole Indian Wars as well as the Mexican War. He served as a mustering and disbursing officer during the Civil War and later retired to California, where he owned a large ranch.

[26]Major Canby to Major Seawell, March 9, 1850, NARG 393, LR, DP.

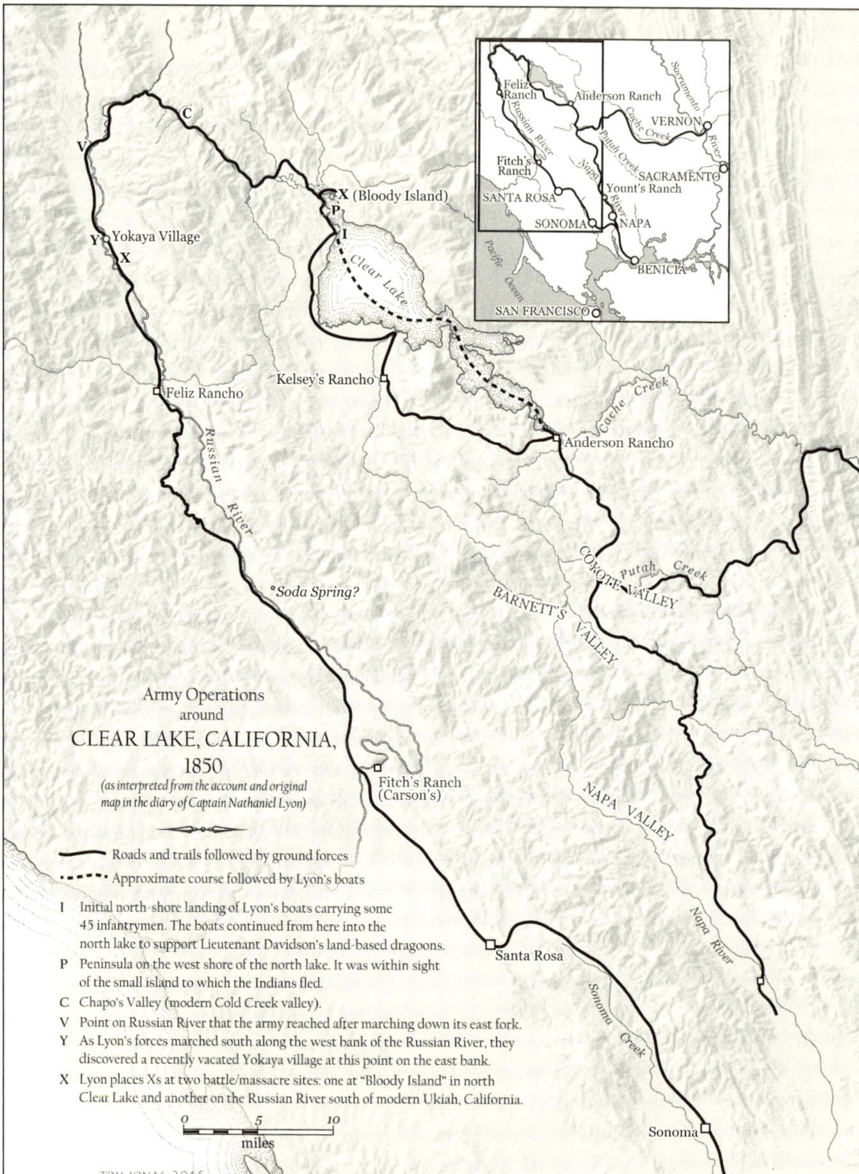

Route of the army expedition undertaken for the "chastisement" of the Goose Lake Indians for killing Captain Warner, May–July 1850. *Map by Tom Jonas.*

tribe on the upper Sacramento that were engaged in the murder of the late Capt. Warner T. E.—Dept. Sep. Order No. 11 a duplicate of which is enclosed herewith, details the companies of which your command will be composed—The rendezvous for this command is left to your discretion, but the Commanding General suggests that it should be assembled at some point between Benecia [Canby consistently misspells Benicia] & Clear Lake, but so far from the scene of your operations that the Indians may not take the alarm before you are prepared to operate against them—Your directions will be given and preparations made so as to commence operations as soon as the roads are practicable. It is supposed that this will be the case in two or at most three weeks after the cessation of the rains, or early in next month—With this object in view a copy of these instructions will be sent to Bvt. Maj. Wessells 2nd Inf (the second in command) with directions to forward the preparations as rapidly as possible in the event of your being absent when the original instructions reach Benicia.[27]

In making your preparations it will be necessary to provide three boats each able to carry 15 men, a sufficient number of wagons to transport your supplies to the immediate vicinity of your operations—wagon floats to facilitate the passage of your wagons across any streams that may be encountered on your route, and the necessary tools for making or repairing roads &c—Pack mules should also be provided to be used when wagons cannot be advantageously employed.

The Commanding General is informed that there is a practicable wagon route from the Napa to the south eastern extremity of Clear lake, by which your boats or your materials for them in readiness to be put together when you reach the lake, may be transported—The same route will probably be the best for your supply train which should follow your operating parties as closely and as far as the roads will permit the passage of wagons.

Your wagon train will be accompanied by a small escort and the remainder of your command be divided into two parties (Infantry and Dragoons) to operate on both sides of the Lake or as you may find most advantageous and as far as possible in concert—It is expected that

---

[27]Just why Lyon was given command of the expedition, rather than Major Henry Walton Wessells, who was mentioned above, is not clear.

the Indians of the lake when driven from their Rancherias will take refuge upon the Islands. The boats will be provided for the purpose of securing such of them as succeed in escaping from the land attack.

The following extract from the instructions of the Major General Commanding the Division will govern you in your operations against these Indians. "The General desires that the commanding officer be instructed to ascertain *positively* the tribe that is guilty," which in regard to the murder, at Clear lake is not difficult as there are witnesses at hand—to attack them wherever he may find them and to pursue them as long as he is able without proposing or attempting a parley or any communication with them—to inflict such a blow as will intimidate them and others, and prevent a repetition of like offenses—to prefer individuals to escape to their capture as prisoners—and not on any account to allow any citizens or others who may be present to take and bring away prisoners. This will only be pretence for procuring servants about their ranchos.

I enclose an extract from a report from Lieut Davidson 1st Dragoons which may give you some useful information in relation to the country in which you are about to operate.[28]

Your command after the completion of this duty will proceed to the Upper Sacramento for the purpose of punishing the tribes by which Capt. Warner was murdered and your preparations will be made with reference to this and to the fact that General Riley has been informed by the Major General commanding the Division that supplies of subsistence &c have already been forwarded to the camp on Bear Creek from which place your supplies for that expedition will be drawn.[29]

---

[28]Lieutenant J. W. Davidson came to California with General Stephen Watts Kearny's command in 1846. Davidson had explored the region around Sonoma in the years that followed, leading a patrol to the western shore of the lake in December 1849. Davidson visited the ruined ranch of Kelsey and Stone, after their murders, and captured several Indians. Others fled before him to the lake. In his report, he suggested the use of three boats to attack the Indians on the island, an idea adopted by the high command. Once on the island, Davidson thought the Indians would have to make a stand and not disappear into the surrounding hills. In 1854 Davidson and his command of sixty dragoons were nearly annihilated by two hundred Jicarilla Apaches southwest of Taos, New Mexico. Davidson later served in the Civil War as a brigadier general. The "extract" that Canby refers to is Davidson to Canby, January 6, 1850, NARG 393, LR, DP. For an extensive biography of Davidson, see Wilson and Fiske, *Appletons' Cyclopaedia of American Biography*, 2: 87.

[29]Bear Creek is a tributary of the Feather River, entering it near Nicolaus, California, the town later named for an early rancho in that region.

The commanding General is unable to give you any definite information with regards to the Indians engaged in the attack upon Capt. Warner's party. He supposes that you will be able to get this information from Col. Casey 2nd Inf. or Lieut. Williamson T. E. You will be guided by the instructions heretofore given you and by your own discretion and the Commanding General confidently expects from your zeal and experience that you will effect much towards the accomplishment of the objects of the expedition: the infliction of such punishment upon the guilty tribes as will effectively deter others from the commission of similar outrages.

You are authorized to supply one or more guides as you may find it necessary and an interpreter for the purposes of communicating with friendly tribes and gaining information &c. &c.

You will make regular monthly returns of your command and frequent reports of your operation, and you are desired to collect any information that may be useful with regard to the country through which you pass, particularly with regard to the number of the Indians, their habits, haunts, and feelings towards the whites, the best positions for the establishment of posts for the purpose of controlling them, the routes by which supplies may be sent &c. &c.

The Commanding General regrets that he cannot detail a Topographical Engineer to accompany your command, and he desires that you will direct some of the subalterns of your command to perform such of these duties as he can, without interfering with his more appropriate duties. A sketch of your route should accompany your reports.

Very Respectfully
Your Obdt. Svt.
E. R. S. Canby
A. A. G.

Major W. Seawell
2nd Infantry
Benecia, Cal.

Just why Major Seawell failed to lead the expedition into Clear Lake is unclear. He may have been incapacitated. Seawell assigned the duty to Captain Lyon, an Irishman who was known to be aggressive and uncompromising. Lyon later commanded the arsenal at St. Louis as the Civil War broke out, defended it against a mob bent on taking it by firing into the mass of people, and was then given command of northern Missouri. During the early weeks of the war, he refused to obey the governor, who sympathized with the South. Lyon raised troops to defend St. Louis and moved on a much larger Confederate force at Wilson's Creek in August. Outnumbered two to one, Lyon attacked General Benjamin McCulloch's army, was shot from his horse, and wounded in both the leg and head. Refusing to retreat, he commandeered another horse, led the final assault, and received a mortal wound in the chest. His efforts saved Missouri for the Union as the Confederate force withdrew.

Lyon's two reports of the assault on the Pomo Indians at Clear Lake are here reproduced. His reports offers considerably more detail about the fighting than previously known. Indeed, much of the information on the so-called Massacre at Bloody Island comes from oral history provided by a six-year-old Pomo girl named Lucy Moore, who survived. She hid in the tule grass as many of her people were being killed and later provided an account of the slaughter. While the oral history certainly is important, and it does not contradict Lyon's official report, a much better understanding of the attack can be gleaned from Lyon's report. The report also demonstrates that merely fifty warriors met him on the field of battle, a much smaller number than often assumed. Indeed, Lyon was so successful because of the small number of Indians who opposed him.

There was a history of trouble near Clear Lake. Violence erupted in 1848 and again in late 1849. The army sent a patrol of forty troops under Captain J. S. Bracket into the region in May 1848. Major James Hardie, an officer of considerable restraint who later served all across the frontier, especially in Texas, arrived on June 1 at the foot of the lake to negotiate a treaty of peace. Some ten chiefs from the region attended, smoked, and ate while Hardie convinced them to remain at peace. Some were likely Wappo Indians who inhabited Napa and Pope's valleys, however the larger number were clearly Pomo Indians from Clear Lake. The Pomos also had villages in the Russian River valley to the west of Clear Lake, although it is difficult to determine whether they were involved in the negotiations.[30] Some of their men, under a chief named Wos, had threatened Henry Anderson's father (of Anderson's ranch) with bows and arrows, and Charles Kelsey had also been threatened. Kelsey and some twenty other citizens, most of them from Sonoma, had filed a petition asking for military protection.[31]

Hardie spent several days surveying the situation and determined that the Indians had grievances. They were used to receiving presents from the settlers, since they believed that the white ranchers had settled on their lands. The ranchers also wanted the Indians to work for them. "I believe," Hardie reported, "that secret efforts [have been] made to get the Indians to work and to this end that stories have been circulated among them by plotters [Californios?] to operate on their fears." Later evidence suggested that Californios may have been

[30]James Hardie examined reports that the Pomo Indians around the lake were "hostile" in May 1848. He concluded that the rumors had "little foundation in fact." Hardie returned in June to negotiate with them. He identifies their leading chief as one Mensia (likely Menaca). There were twelve rancherias in the vicinity of the lake and to the south of it with about fifty Indians in each of them. Obviously, by the time Lyon appeared with his force two years later, those numbers had dropped, perhaps by as much as half (Lyon seems to indicate that he faced only about fifty warriors in the final battle). Those on the lake were Pomo Indians, while the others, farther south in the Napa valley, were Wappo people. Hardie is unaware of the difference, and culturally there was very little difference. He did report that "the terrible scourges [malaria?] and chastisements of the days of former governments [Mexican Californios] have effectively cured them of any disposition . . . to provoke hostilities." A party of Mexican Californios had kidnapped a number of Indians—likely friendly Wappo Indians from Napa—the year before. When the Indians fled, the Mexicans killed several. See Major E. L. Bale to the Commanding Officer at Sonoma, March 17, 1847; Hardie to Lieutenant William Tecumseh Sherman, May 28, 1848; and Hardie to Governor Richard B. Mason, June 1, 1848, NARG 393, LR, 10th Military Department. See also Kroeber, *Handbook of California Indians*, 217–39; Jesse O. Sawyer, "Wappo," 256–63, Sally McLendon and Robert L. Oswalt, "Pomo: Introduction," 274–88, and Sally McLendon and Michael J. Lowy, "Eastern Pomo and Southeastern Pomo," 306–23, all in Heizer, ed., *California*.

[31]See Petition and Testimony of Charles Kelsey, May 22, 1848, NARG 393, LR, 10th Military Department.

encouraging Native resistance to the expansion of Anglo ranchers northward. Chief Wos, who refused to negotiate with Hardie, especially complained that he did not want to work for white ranchers.[32] Over the next two years, the issue of work seemed to escalate into conflict.

The most controversial part of the tragic events at Clear Lake is the depiction that emerged in the oral history relative to the two ranchers who were killed. Andrew Kelsey and Charles Stone have been depicted as slave hunters in the oral history. They supposedly forced Pomo men to work for them as miners and then refused to pay them for their work, a number of them dying as a result. Kelsey also apparently demanded and took young Pomo girls for use as sexual servants. When he supposedly abducted the wife of Chief Augustine as such a servant, the oral account suggests that she rendered the gunpowder in the ranch house useless by pouring water on it. The Indians then attacked the ranch house and killed the two men, prompting Lyon's retaliatory expedition.[33]

There is no evidence from Lyon's report that he knew of the earlier actions of Kelsey or Stone, but he likely should have realized that Indians in California were being oppressed by miners and ranchers. It is certainly true that Californios and early Americans often kidnapped Indians for use as servants. On the other hand, other ranchers in the region, including several Mexicans as well as the Anderson family, were left alone. On another issue, the oral account also suggests that the Indians responsible for the attack on the ranch house, Augustine's band, were not on the island when it was attacked—that Lyon chastised innocent Indians. Lyon's report does shed light on the issue of guilt, suggesting that the truly guilty parties came from the Russian River. Although the Pomo Indians at the island did put up a stiff resistance, their arrows were nevertheless ineffective.

[32]See Major James Hardie to Col. R. B. Mason, June 1, 1848, ibid.

[33]The oral account is found in several places. See Elizabeth Larson, "Bloody Island Atrocity Remembered at Saturday Ceremony," *Lake Country News* (Lakeport, Calif.), May 13, 2007.

Captain Lyon's First Report on the Clear Lake Expedition

Head Quarters
Clear Lake Expedition
Anderson's Ranch[34]
May 22, 1850

Sir:

In compliance with Department orders (Special) No.24, I proceeded from Monterey to Benicia, where I arrived on the night of the 4th inst. And the next morning took command of the Expedition designed to proceed against the Indians on Clear Lake & Pit River, by virtue of Majr. Seawell's order of that date (a copy of which is herewith enclosed), and setting out next day (6th) from Benicia, I reached this position at the south end of Clear Lake on the 11th. The next day the Dragoon Company (Lieut Davidson) was detailed round the Western shores of the Lake to cooperate with the Infantry, to proceed by water up the lake. The Indians on learning our approach fled to an island at the northern extremity of the Lake, opposite to which and on the western shore of the lake, the command took position on the afternoon of the 14th, the Indians still gathering rapidly on the Island. Lieut. Davidson with Lieut. Haynes (mountain howitzer) attacked a Rancheria on the morning of this day, killing four & securing an Indian chief.[35] Early on the morning of the 16th the two shores being guarded, the landing on the Island was effected under a strong opposition from the Indians, who perceiving us once upon their island took flight in every direction, plunging into the water among the heavy growth of Tula which surrounded the Island, and which on the Eastern

[34]Lyon's first report is found at Lyon to Canby, May 22, 1850, NARG 393, LR, DP. Some confusion exists regarding Anderson Ranch, which is today a state park. Apparently, Henry Anderson started the ranch sometime prior to 1850, but it was then sold to John Grigsby in 1854, who in turn sold it to John Still Anderson, a Scotsman who arrived on the scene in 1884. Today the park is named for the latter Anderson rather than the original owner of the ranch. For reference to Henry Anderson, the first owner, see L. W. Boggs to Governor Mason, May 22, 1848, NARG 393, LR, 10th Military Department.

[35]The chief is not identified. Lieutenant Milton A. Haynes, a graduate of West Point in 1838, served in the artillery.

& Northwestern sides extends to the shores. Having rapidly cleared the Island, I saw no alternative but to pursue them into the Tula, & accordingly orders were given that the ammunition be slung around the necks of the men and [that] they proceed into the Tula and pursue and destroy as far as possible; the tula was thus thoroughly searched with severe and protracted efforts, and with most gratifying results—the number killed I confidently report at not less than sixty & doubt little, extending to a hundred & upwards. The Indians were supposed to be in numbers about 400—their fire upon us was not effective, and no injury to the command occurred.[36] The Rancheria extending about half way around the island was burned, together with a large amount of stores collected in them.

Being satisfied that the Indian tribes on Russian River had participated in the murders of Stone & Kelsey, and were now harboring one or two tribes known to be the most guilty, I now proceeded to the headwaters of that River, seeking first a tribe whose chief is called Chapo; but finding the Rancheria deserted to which my guide led me as his, I caused a thorough but ineffectual search to be made in the vicinity, and then proceeded down the river for about 22 miles to a tribe called the *Yohaiyahs,* among whom was Priesta and his tribe—the most active participants in the atrocious murders—I found them early on the morning of the 19th on an Island formed by a slough from Russian River, which was covered with dense undergrowth, & in the part where the Indians were mostly concealed, were many trees both dead and alive in a horizontal position, interwoven with a heavy growth of vines.[37] Their position being entirely surrounded they were attacked under most embarrassing circumstances, but as they could not escape, the Island soon became a perfect slaughter pen, as they continued to fight with great resolution and vigor till every jungle was routed—their numbers killed I confidently report at not less than seventy five, and

[36]Note that Lyon later reports that only fifty arrows were launched at the soldiers in the initial attack by the Indians, suggesting that he engaged a mere fifty Indians.

[37]Neither Priesta nor Chapo is mentioned in the oral history accounts. The oral history account suggests that the killing of the ranchers was effected by Chief Augustine's people. Chapo's village is located on the map that is placed with Lyon's second report, which follows, as position *C.*

have little doubt it extended to nearly double that number. I estimated their whole number as somewhat greater than those on the Island before mentioned.[38] They were bold and confident, making known their position in shouts of encouragement to their men, & defiance to us. Two of their shots took effect wounding somewhat severely Corporal Kerry & Pvt. Patrick Coughlin Company "G" the former in the shoulder & the latter in the thigh.

A body of Indians supposed to have been concerned in the outrages at Kelley's Rancho, & who it was believed were harboring one of the tribes known to have been concerned in the Kelsey murder, lay about 10 miles below; & in order that action might promptly be taken against them according to the circumstances in which they might be found, I detached Lieut. Davidson with his (Dragoon) Company to proceed hastily to the spot so as to anticipate an alarm from the events just mentioned, and obtaining with the assistance of Fernando Feliz upon whose land those Indians lived, the facts, he was instructed to act accordingly; on arriving at Fernando Feliz Rancho he found the Indians had fled through fear, tho' the intelligence that the hostile tribe was harbored by them proved unfounded, and no definite intelligence that they had participated in the murder aforesaid was ascertained.[39] During our passage down Russian River an Indian was taken captive, who communicated some very unexpected intelligence that

[38]Lyon's population numbers give 400 Pomo Indians living near the lake and about 500 on Russian River. Along with the few friendly Indians living to the south of the lake, the total population was likely fewer than 1,300. Wessells, who came through the region in the next year (his journal is to follow), noted that estimates found in newspapers were outrageously high. Kroeber, who used some of these sources, suggests that each of the seven Pomo groups had roughly 1,000 people, or 2,000 for the Clear Lake and Russian River Indians combined. He also gives a figure of 1,000 for the Wappo, making 3,000 in all for the region invaded by Lyon. These estimates are more than double the real populations, at least in 1850. See Kroeber, *Handbook of the Indians of California,* 221–32. The articles on the Pomo Indians in Heizer, ed., *California,* make no real attempt to ascertain population changes from disease, which are apparent just between 1848, when Hardie visited the region, and 1850, when Lyon came onto the scene.

[39]Fernando de la Trinidad Feliz received a 17,700-acre Mexican land grant in 1844 in the heart of the Pomo Indian country. The Rancho was one of two granted in the Russian River, where the Indians still controlled the land. The ranch was near Hopland, California. See Ogden Hoffman, *Reports of Land Cases Determined in the United States District Court of the Northern District of California* (San Francisco: Numa Hubert, 1862), 69–71.

some citizens (Spanish) had instigated the Indians against Americans, confirming in this respect the hints previously thrown out to me by several persons. Lieut Davidson informed me that if this statement were true, the evidence of it must be found among the agents of these individuals (Spaniards), and that these agents were living on the road to Sonoma and at that place. I accordingly detached Lieut Davidson to proceed to Benicia by way of Sonoma, taking with him the wounded & charged to obtain all the information possible upon the subject above mentioned, & to act accordingly, while the Indian who communicated this intelligence was dispatched with the promise to bring his chief & principal people to the head of the Lake, & meet me tomorrow, & should they meet me and confirm his statement, I shall endeavor to secure enough of them to establish the facts, and send them to safety in Benicia.[40]

Leaving the valley of Russian River I proceeded across the mountains dividing the waters of the River from those of the Lake, and after two days march arrived (yesterday about 2 o'clock P.M.) at this place—Submitting respectfully the above brief account, I reserve for a more convenient opportunity a detailed report to be accompanied with a map which I shall furnish at an early day.[41]

I am Sir
Very Respectfully
Your most Obt Servt.
N. Lyon
Bvt Capt 2nd Inftry
Comdy Expedition

Maj. E. R. S. Canby
Asst Adjt Genl
Monterey, Cal

[40] Apparently, the investigation found no appreciable evidence of "Spanish" meddling.
[41] This is a reference to his second, much longer report, which follows.

Captain Lyon's Second Report on the Clear Lake Expedition

Head Quarters
Clear Lake Expedition[42]
Anderson's Rancho
May 23, 1850

Sir:

In obedience to Dept Special Orders No.24 assigning me to such duty with the Expedition designed to operate upon Clear Lake & Pit River as Maj Seawell might assign to me, I proceeded to Benicia & reporting to Maj. Seawell was assigned to the command of Co. "G" 2nd Infy & being the senior officer on duty with the expedition, the command of it devolved upon me as shown by Maj. Seawell's order of the 5th Inst. (a copy of which has already been forwarded), & receiving at the same time from Maj. Seawell the announcement that all things were in readiness for the movement of the Expedition, it left Benicia the next morning and moved that day (6th) to near Lousquall's Rancho, a distance of 15 miles, thence the next day to Younts Rancho, marching about the same distance as the day before.[43] There I found Lt. Davidson with his Co. ("G" 1st Dragoons) awaiting our advance. From Younts to the hills at the head of Napa Valley (along which our road lay,) is about ten miles distance, making about 45 miles from Benicia, in a north westerly direction, of a very good practical wagon road excepting several gullies, the depth & abruptness of which rendered them difficult of passage with loaded wagons.

Having proceded [*sic*] on the 3d day (8th) about two miles into the hills, I came to a long and very steep hill, (having at least 45° elevation) over which we straggled for a long time, by doubling & trebling

[42]This second, more detailed report is found at Lyon to Canby, May 23, 1850, NARG 393, LR, DP. It included a map with various designations on it, such as *I*, *Y*, and *P*, which Lyon used in his report to identify locations.

[43]George Calvert Yount was a trapper who arrived in California with William Wolfskill in 1831. He received two Mexican land grants in the Napa valley in 1836, totaling roughly 16,000 acres. Lousquall's rancho was close to present-day Napa, California. See Hoover, Rensch, and Rensch, *Historic Spots in California*, 239–40; Ellen Lamont Wood, *George Yount: The Kindly Host of Caymus Ranch* (San Francisco: Grabhorn Press, 1941).

teams, to pass only a portion of our wagons when convinced that if further obstacles of the kind were to be overcome, it would be vain to attempt farther advance with the wagons, & accordingly on examining the road for a few miles in advance, I had no alternative but to abandon the wagons & the work of preparing for our farther advance by means of pack mules was at once commenced. On packing the next morning, our Pack Saddles & fixtures were found to be sadly defective,—the saddles being much too wide, by the sides forming more than a right angle with each other, and the front and hind parts of the same dimensions the saddle in position naturally tipped forward on the mule, & being short, but slight movement displaced the entire pack,—the width and unsuitableness of the saddle to fit the mule making it impossible to secure it upon the back, while all our lashings provided were of stiffly tanned rope, with which it was utterly impossible to have secured a pack upon the best of saddles, while all the saddle rigging (of which a Breeching & Breast strap were needed) consisted of a crupper, so thin & narrow as to have cut and galled the mule if of sufficient strength, but being of slight sheep skin, broke upon the first use, while the fastenings of straps (extremities of the firth) to the saddle served in like manner, added to this our mules were mostly wild & our men unacquainted with packing; so that by the above account you will perceive sufficient aggravation, without that of finding our saddles made of a kind of rotten oak, which broke as ready as if made of clay; & one indescribable scene of confusion made up the days occupation, in which officers, enlisted & Quartermaster's men toiled with most laudable zeal;—our Boats too, provided with imperfect carriages, necessarily because the baggage wagons of more than half our train marched about 6 miles to Barnett Valley, passing into it down a nearly perpendicular steep down which our best carriages were let with ropes.[44] This valley has great extent & fertility & an opening to the east, extending through windings among the hills to the east to Sacramento valley. Marched next day (10th) to Coyote

[44]The thin saddle rigging described as a "crupper" was actually a strap looped under the mule's tail to keep the pack saddle from sliding forward. Barnett valley began near the present town of St. Helena, today a major wine-producing region north of Napa.

Valley, a distance of about 10 miles passing over a very steep rough hill about 1½ miles long, where it became necessary to double teams and proceed very slowly with the boats.[45]

Marched next day to Anderson's Rancho, at the southern extremity of the lake, distance about 18 miles, over a tolerably good wagon road. Subsequent to the outrages at Kelsey's Rancho, Lt. Davidson had secured and retained an Indian belonging to the guilty tribes whom I found in his possession, & who proved to be of most essential use, possessing great intelligence & a knowledge of all the languages spoken by the different tribes with whom we met.[46] This account together with those of two Indians whom Lt. Davidson procured at Popes Rancho in Barnetts valley, acquitted the Indians at the southern extremity of the lake [Wappo Indians] of participating in the murder of Kelsey & Stone, or of sympathizing with the murderers.[47] Mr. Moses Carson, who generously volunteered his services as guide, adding that the Tribe or Tribes at the South end of the Lake were not only friendly to the whites, but at war with those concerned in the outrages at Kelsey's Rancho.[48] On communicating with these Indians I became satisfied of the correctness of the above accounts, & their accounts agreeing with what intelligence I had previously gathered of the feelings & position of the tribes to the north, conveyed to me the following impressions, upon which I acted, and to a proper understanding of which, I invite

[45]Coyote valley is farther northwest of Barnett valley.

[46]This Indian had apparently been captured by Davidson in December 1849 and brought to Sonoma for questioning.

[47]William Julian Pope received a grant for Rancho Locoallomi in 1841, in what became known as Pope's valley. He died in 1843, and his widow then married Elias Barnett, whose name was attached to Barnett's valley. The innocence of the Wappo Indians on the south end of the lake corresponds with the reports of Lieutenant Hardie, who negotiated with them in 1848. For Rancho Locoallomi, see Hoffman, *Reports of Land Cases,* 141–42.

[48]Moses B. Carson was then working on Henry D. Fitch's rancho called Sotoyome, located near Healdsburg, California. Carson was an experienced frontiersman, being the older brother of Christopher "Kit" Carson. Moses had apparently entered California with the Bartleson-Bidwell party, the first groups to attempt taking wagons across the deserts of Utah and Nevada into California in 1841. He then went to work for Fitch but, after Fitch died, sold a large herd of cattle from the ranch and departed for the East soon after the Lyon expedition. The sale produced a large court case, *Fitch v. Brockman,* where the issue was clearly relative to who actually owned the cattle. See *Reports of Cases Determined in the Supreme Court of the State of California,* vol. 3 (San Francisco: Bancroft-Whitney, 1888), 348–75.

attention to the accompanying map of our route & operations drawn by Lieutenant Haynes.[49] That the different tribes concerned in the outrages upon the Lake, encouraged by their success in escaping the pursuit of our troops last winter, were resolved on war and were on the day of our arrival at the Lake (the 11th) to have a general gathering at Kelsey's Rancho or its vicinity, & killing and feasting upon the cattle there, were to celebrate & confirm their resolutions, & make this position their advance post for meeting their enemy, because of its advantages to observe from the surrounding hills all that might transpire for some distance around, & in case of repulse from this point, were to retreat to a fort some distance to the north & near the Lake, which had been prepared for strong defense, & injuring horsemen, who, (from their previous experience), were alone expected to approach, opposition to this position was represented by an island in the Lake, easy of access to them and to which they would take refuge in the last resort.[50] And finally, that all the hostile Indians were to the west of the Lakes, & none on the eastern shore, my object was therefore to attack & dislodge them at or in the vicinity of Kelsey's, & pursue with haste & attack & drive them from the fort, giving them every possibility for escaping to the Island, which being guarded from the shores, would be finally reached by the Boats.

To effect this, Lt. Davidson with his Dragoon Company was dispatched (on the morning of the 12th) to proceed round on the western side of the Lake, with directions to get position in the vicinity & to the west of Kelsey's Rancho on the night of the next day, so as to advance & cooperate the following morning at 4 O'Clock with the Infantry, who were to proceed during the same night by the Boats to the vicinity of the position, and landing yet in position to the south, of an old Rancheria, & supposed present location of the Indians, to the east of Kelsey's & near the lake Shore. This design was entirely executed, the cooperation being complete, & the entire ground about Kelsey's (tho thickly wooded) for some distance was thoroughly examined, in an

---

49 While not a topographical engineer, Haynes had undoubtedly studied mapmaking at West Point.

50 Clearly, Lieutenant Davidson and his dragoons, who had campaigned in the region south of Kelsey's ranch in December 1849, are being referred to in the report.

ineffectual search for Indians.[51] Our intelligence proved incorrect,—no other signs of Indians being discovered except that of a few who had been there a few days previous to procure beeves. Being reassured by my Indian informants of the certainty of a strong position at the fort to the north, and opposite the Island at the head of the lake, the Dragoons were directed to proceed to it as soon as possible, with a view to cut off escape except to the Island, while the Infantry party was to continue by water to a convenient vicinity, to land and join in the attack upon the Fort. Lt. Davidson in advancing, caught an Indian, who afterwards turned out a chief, & who conducted him to a hostile Rancheria, secure enough by nature and the accidental fall of trees about it, to be called a Fort, & is believed to be the one referred to by the accounts above mentioned. The account & result of the attack upon this point will be seen in Lt. Davidson's report of it herewith enclosed.[52]

Proceeding with the Infantry party up the Lake at this time, on hearing the discharge of (Lt. Hayne's) artillery, I immediately effected a landing at the point *I* (see map), with a view to cooperate with Lt. Davidson, or attack those that might escape from his position.[53] Lt. Davis was detached with a small party to move south westerly & near the lake, while I proceeded accompanied by Lt. Paine to the west meeting and pursuing four Indians, & crossing the path upon which Lt. Davidson soon after followed the Indians to their Island, as seen by his report.[54] I proceeded considerable distance interior, in an unsuccessful attempt to meet Lt. Davidson, & on returning to the Boats found he had followed the Indians to their Island, and taken

[51]Lyon sent Davidson to attack the supposed stronghold of the Pomo Indians just north of Kelsey's ranch. Lieutenant Haynes, in drawing the map of the movement, noted an Indian village identified as Yockaya, just above which he put crossed swords, indicating where the engagement was supposed to have occurred. The "Yockayas" were undoubtedly the Pomo band that Lyon later identifies as "Yohaiyahs." They are designated with a *Y* on the map, which is where Davidson expected to find them. When he arrived, they had fled farther up the Russian River, and the army turned instead to Pomo Indians nearer Clear Lake. Another phonetic form is Yobakeya, which is used in McLendon and Oswalt, "Pomo: Introduction," 282–83.

[52]Davidson's official report is missing from the War Department records.

[53]Point *I* is difficult to ascertain on the map. It shows up in the lake, on its western shore, about eight to ten miles below the island where the final assault occurred.

[54]Lieutenant Nelson H. Davis graduated from West Point in 1846 and saw service during the Civil War as an inspector general. Lieutenant F. Paine apparently resigned from the service shortly thereafter.

position opposite to them & secure from their view behind a hill at the point *P.*[55] I here joined him with the Inf'y leaving Lt. Davis with the Boats to bring them up after dark. The Indians now fully alarmed & not yet aware of the presence of our boats, were rapidly collecting from all directions upon the Island, and confident of safety there, howled the long night, in fierce denunciation, & bold defiance, of us, encouraging at the same one another in their hostile attitude. The Island being nearer the eastern shore & having a wide sheet of water to the west of it, & thinking it impracticable from the extent of the tule to the north to pass round it, the design was, to land the Dragoon party during the night on the opposite shore, in time to get position to guard it, by daylight, when the Inf'y party were to proceed to the Island, to make the attack, leaving a small party in the Boats, which were to cast off & take position to the west & north to cut off all attempts at escape in those directions, by means of the Tula Boats. The attempt to land the Dragoons upon the opposite shore proving unsuccessful in consequence of the heavy growth of Tula, the Dragoon party was directed to proceed with horse around the head of the Lake if possible, & get the position before intended,—This order was executed with the greatest promptness & most gratifying success, causing but little delay.

The original design was now executed—Dr. Dyenle being left in charge at camp, & Lt. Haynes charged with the care of the Boats after the attacking party should effect a landing. Giving their attention to the Dragoon party as it took position, the Indians taunted them in the boldest terms of defiance, challenging particularly Mr. Kelsey & his son who were with us, alluding to their outrage upon his victimised [*sic*] brother, in most exulting manner, till the approach of our boats gave their attention more absorbing occupation; & to meet which the warriors advanced & took position, upon the hill, which forms a little promontory on the south side of the Island, & behind the crest of this hill they concealed themselves watching our advance, till thinking us sufficiently near, they launched forth their arrows about 50 in number, taking the precaution to elevate them well to secure a great range,—some passed over us, a few fell among us without effect, but

[55]Point *P* is difficult to make out on the map, but it appears to be located on a hill, due south of the island, just across the water from it.

most fell short of us,—a stone was slung with great energy, & passing over us fell with violence into the water,—another of less weight fell short of us.[56] In advancing upon the west side of this promontory, the Indians kept their position behind the crest of it on the east, & Lt Paine was directed to move with his Boat to take them in flank, which being done a landing was easily effected; the Indians taking flight in utmost haste, pursued by our troops to the north & east sides of the Island, & plunging into the water among a heavy growth of Tula, intending concealment among it, & their numerous Tula Boats, & thereby final escape. Having no alternative but to pursue them into the Tula, I directed the ammunition to be slung around the necks of the men, & they proceed immediately into the Tula in pursuit of the fugitive enemy. This arduous work was fully participated by officers & men, & cheerfully and effectively executed in a protracted & thorough examination of the Tula to the shores, & with most successful results; Lt. Davis killing two warriors with his own hands; Lt. Davidson dismounted his company, and advancing, with most successful results, into the Tula thoroughly scoured the portion of it to the east of a deep stream, passing to the east of the island, & which separated his party from that of the Inf'y; upon the banks of which & towards the northern extent of the Tula, near the close of operations, I met Lt. Stoneman enthusiastically engaged in this arduous work.[57] Lt. Haynes charged with the Boats, did effective service in cutting off attempts at escape, & detecting & destroying their numerous Tula Boats, & Burning the extensive Rancheria & collection of stores.

The lake does not exceed 40 miles in length, & my impressions of its contour, (having passed through its entire length three times, while Lt. Haynes saw it only from the shore) differ somewhat from those given in the map by Lt. Haynes; this difference is indicated in pencil mark, there being a large bay in which are two islands, setting down to the south east of the long point, projecting from the west about the centre

---

[56] The use of stone slingshots in the skirmish was unusual, although perfectly understandable. The Kelsey referred to was Charles Kelsey.

[57] Lieutenant George Stoneman graduated from West Point in 1846, having had Thomas J. "Stonewall" Jackson as a roommate. He served in California for several years and later in Texas. During the Civil War, he served as a general officer, mostly commanding cavalry units. Stoneman later returned to California and was elected the fifteenth governor of the state in 1883.

of the lake. The island put down as east of that point is south of it, & somewhat to the west of its eastern extremity. The Tribes occupying the lake north of Kelsey's Rancho, combined in the outrages at that place, & in determined hostilities to the Americans, & who were united & engaged in the action above described, were seven in number, several of whom had their Rancherias on the Island, while several Rancherias like that Lt. Davidson attacked, were said to be located in the nooks & ravines among the mountains which surround the lake.[58] These Rancherias I felt desirous of seeking, and should not have done so, but that our recent affair, together with a subsequent advance among the hills, would cause alarm to the tribes on Russian River, who gaining time by such delay would take flight & effectually evade our approach. But being yet in considerable numbers, & fired with the same hostile spirit as those at the lake, in whose outrages I was satisfied they had participated, & in whose feelings they now sympathized, I deemed an effective chastisement of *them* of more importance than a continuation of that already severely felt by those on the lake.

Remaining therefore one day in camp to obtain provisions from our Depot at Anderson's Rancho, (at the South end of the Lake) by means of our Boats, the command set out early on the morning of the 17th by marching about half through the mountains which separate the waters of the Lake from the head waters of Russian River. I halted till sunset, when I moved into what I will call Chapo's valley to a point *C*; Chapo being the chief & designation of the Tribe occupying this valley, and whose Rancheria was represented by my guide, as being on a small stream directly in front of my position at C, & which I designed attacking at day light next morning, & for the execution of which, I moved accordingly, & after much seeking found the deserted Rancheria which this tribe formerly occupied, & continuing what I believed a thorough but ineffectual search, separating the Inf'y & Dragoon parties, (as will be seen by the lines drawn) but finding

[58]The reference to seven rancherias fits well with Major Hardie's account of ten in 1848, three of those identified by Hardie being Wappo Indian rancherias. Hardie puts the population of two of these towns as just fifty people, but they were likely somewhat larger in Pomo country. While Lyon gives the population of the Indians he attacked on the lake as 400 in his first report, if Hardie's figure of 50 people per rancheria is somewhat accurate, then the seven "tribes" mentioned by Lyon would consist of roughly 350 people, quite in line with Lyon's overall estimate.

no signs of Indian habitations, the movement was continued during the day (18th) down a tributary to Russian River to a point *V*, in the direction of a tribe called Yohaiyahs, who it was well ascertained had participated in the Kelsey outrage, & were now harboring Prietia & his tribe of Habinapos, the boldest & most guilty in the murders of Kelsey & Stone.[59] Advancing the next morning (19th) about six miles we came upon the Rancheria of the Yohaiyahs, now vacant, but with every appearance of having been left but a few minutes before, crossing the stream to the east at this point, we found a dense hammock on our left, which Lt. Davidson was directed to examine on the eastern skirt with his Dragoon Company, & halting to direct the Inf'y (just in rear) to some point of examination, I perceived them in full force before me, in the hammock by the shouts of defiance to us, & encouragement to their men raised on perceiving Lt. Davidson's line extending along the high bluff, on the edge of the wood to their front, where he halted to guard their escape on that side, while the Inf'y party were taken back to get position on the peninsula, formed by the stream & the slough from the stream the latter forming the Island upon which the Indians were. The line of infantry being formed (facing east and deployed as skirmishers) it was directed forward to cross the slough, & attack the enemy on the Island, but upon approaching it, it was found impassable, tho the skirmishers opened a somewhat effective fire upon the Indians concealed in a dense jungle on the opposite side, from which they shot many arrows. Making to the right, I effected a crossing with some difficulty accompanied by six men, the remainder having gone to the left to seek a crossing. I immediately found the Indians in front too numerous for my small party, which was now within effective range of their fire, by which two were soon disabled, tho, the dense jungle through which their fire was directed, served to

[59]Lyon seems to place the Habinapos as inhabiting the Russian River valley. They may have been related to the rancheria that Lyon designates as Chapo's, but that is unclear. It does seem obvious that Indians from Yockaya rancheria (designated on the map as *Y*) and Chapo's rancheria (designated as *C*) had fled in front of Davidson's cavalry and were then in the headwaters of the Russian River, at a point designated *V*. It is impossible to determine from the map the exact location of point *V*, but it appears to be north of Chapo's village, near the headwaters of the Russian River. McLendon and Oswalt have little knowledge of the Habinapos, other than to speculate that their rancheria was in Kelsey Creek. See "Pomo: Introduction," 287.

divert it, & weaken its force. This small party continued it exertions with most gratifying effect, & was promptly relieved, (from what for a few minutes seemed a delicate position), by the cooperation on my left of Lieuts. Davis and Paine, who having effected a crossing, directed a destructive fire upon the Indians in my front, against their flank. Flight would doubtless now have been their resort, but finding themselves surrounded on all sides, many continued fighting with great bravery, others seeking escape by concealment, either among the dense undergrowth, or by plunging into the mud or water of the slough whenever they could reach it,—nothing now remained but to hunt out & destroy such of their warriors as remained concealed, care being taken not to injure the other members of the Tribes, but this becoming a work of slaughter *ad libitum* [at ones' pleasure], was soon stopped, & the troops on the Island recalled, those guarding it, remaining in position to destroy fugitives the Dragoon Company having been effective in this service during the entire operation. Lt. Haynes was also occupied in part in guarding the Rancheria, & the large supply of stores collected, which duty was promptly executed. Dr. Deyerle, in addition to prompt attention to prof' [professional] duty, here as on all occasions, seeking to render service, contributed to the harmony of the cooperations [*sic*] by communicating from me to the other portions of our line, tho exposed to a cross fire in doing so.

The numbers killed here, I believe to have somewhat exceeded a hundred, & report the minimum of seventy five, the offensive chief Priesta, was believed to have been killed.[60] Previous to leaving this point, several parties were sent back upon the Island at different times, who found warriors attempting to escape, and about twenty of them were thus destroyed. No other casualties happened to our party than the two wounded before mentioned. About ten miles below here, at Fernando Feliz's Rancho, was a party of Indians, faintly pointed at by my sources of information, as concerned in the outrages on the Lake,

[60]There is a certain irony that the "Massacre at Bloody Island" is relatively well known by the Pomo Indians residing in the region today, as a result of the oral history, but that this attack on Russian River, where even more Indians were slaughtered, has been given very little attention by California historians. The number killed is certainly significant, and may account for the conclusion of McLendon and Oswalt that, at least for the Yobakeya (Lyon's Yohaigah), "these people were never numerous and have long been extinct." See "Pomo: Introduction," 282.

and now harbouring [*sic*] Monaca [Menaca, as reported by Hardie], a chief known as among the guilty. I order[ed] that these Indians should not take flight from alarm at the events just mentioned before action should be taken according to the circumstances in which they might be found. I dispatched Lt. Davidson with the Dragoon Company to proceed as rapidly as possible to the spot & the result of the execution of this order will be seen by reference to his report. As soon as litters could be fitted up for our wounded, I proceeded to join Lt. Davidson at Feliz's Rancho. Believing my orders in regard to the operations in the vicinity of Clear Lake executed with as desirable effect as circumstances would now allow, I directed next day (May 20th) that Lt. Davidson proceed with his Company taking the wounded in charge, to Benicia, there to advance as far as possible the interests of this command, preparing to its movement upon the Pit River, being also charged with the mission to which I have before referred, in regard to detecting citizens (Spaniards), supposed to have instigated Indians against the Whites.[61] From the same point I proceeded with the Infantry party to Anderson's Rancho in two days march along the line indicated, passing Kelsey's Rancho. I expect to leave this point in the morning tomorrow, for Nicolaus on Feather River, at which place Lt. Davidson is directed to join me preparatory to advancing towards the head waters of Pit River.

That the officers of this command have done their duty under the circumstances here mentioned, it is hardly necessary to remark as the Commanding General will have expected this from their well known character, but it is but justice to them, while it affords me infinite pleasure, to bring to the notice of the General, the cheerfulness with which all hardships incident to such service have been met, every opportunity sought & exertion made to forward the Expedition, tendering each other assistance & deeds of kindness by which the pleasure and harmony, if not the success, of our operations have been advanced. Lt. Paine, the A. G. [adjutant general] & A. Q. M. [assistant quartermaster] to the expedition was enabled to accompany us in our operations by leaving his very competent & efficient assistant & agent in those duties, Mr. C. C. Green, in charge at the Depot at Anderson's Rancho,

[61]No evidence ever emerged that led to information regarding the collusion of Californios in the murders.

from which our supplies were promptly forwarded.[62] The above remarks in regard to the officers, apply equally in the main to Rank & file of the command, tho I regret to say there were some exceptions, inhibiting however in beautiful contrast, the main portion, whose efforts contributed so much towards the accomplishment of our objects, & equaling the utmost wishes. In closing this report, if I be leave to state my impressions, that to secure tranquility to both the Indian & white man, & a peaceful intercourse between them, when living in the vicinity of each other, it will be necessary to provide a man of character, ability, & strict integrity, to become the agent for a suitable number of tribes, to teach them their duties & their rights, to enforce the one & *secure* the other, & who in the legitimate exercise of his duties, would avert much that is apparently & perhaps truly criminal in both parties.

Respectfully Submitted
N. Lyon
Bvt. Capt. 2nd Infantry
Comy Expedition

To
Maj. E. R. S. Canby
Asst. Adjt. Gen'l.
Head Quarters

Captain Lyon Makes Preparations for the Pit River Expedition

Head Quarters
Clear Lake Expedition[63]
Anderson's Ranch
May 22, 1850

Sir:

Aware that the General Commanding Division desires that every facility be afforded for expedition in the movement directed against

[62]Green was apparently a civilian, hired by the army.

[63]The original letter is Lyon to Hooker, May 22, 1850, NARG 393, LR, DP.

the Indians on Pit River, in consequence of his more ready means of action by contiguity to the troops of this Command & to the source of supplies, I have deemed it proper to enclose herewith a copy of a letter of this date addressed to Maj. Canby, setting forth the wants of this command before setting out for the head waters of Pit River. I also beg leave to mention that I have no map of the country over which I am to pass, & on speaking to Lt. Williamson on this subject, he told me that he was about leaving at Benicia, in charge of Mr. Steingrantz (one of the Qr Mr's [quartermaster's] men) two copies of the map of Warner's trail, which he had made for Genl Smith, & which could be found in the Qr Mr's store room in a tin case.[64] Could the Genl lend me one of these copies I doubt not it would prove useful & I will endeavour to preserve & return it to him in safety.

Very Respectfully
Your Obedient Servant
N. Lyon
Bvt Capt. 2nd Inf Comg

To Lt Col. Hooker, A.A.Genl.

Captain Lyon Continues Preparations for the Pit River Expedition

Nicolaus Gold[65]
June 30, 1850

Col:

Having obtained steam transportation for the supplies required for the Pit River Expedition the Brig Adelade which is now in readiness to move will proceed to Vernon & there turn over the supplies to the Steamer Lawrence which is to be there tomorrow evening with a view

---

[64] This reference is to Williamson's map, included here with his journal.

[65] Nicolaus Allgeier built a ferry over the Feather River in 1843 just south of where Bear River joins the Feather. It was a natural stopping place for road traffic between John Sutter's fort and Sutter's Hock Farm. The small town of Nicolaus remains today. See Lyon to Hooker, June 30, 1850, NARG 393, LR, DP; Hoover, Rensch, and Rensch, *Historic Spots in California*, 544.

to start on the next morning & proceed with the stores to Lassen's.[66] The Adelade awaiting at Vernon further orders from Maj. Allen. I have also a prospect of promptly obtaining everything required for the outfit & hope to move from this point soon an event not only desirable with a view to the execution of my Orders but to escape the approaching hot weather & sickly season. Should it be found on advancing as far as Lassen's that there is too much snow or water in the mountains for our operations, I can move safely to delay at that point or higher than here—I should think the 10,000 rations will be sufficient for our use but should our operations be necessarily protracted I shall take the precaution to renew our supplies in time. The road between here & Lassen's is believed to be practicable on both sides.

Very Respectfully &c
N. Lyon
Bvt Captain & Comy

Lt. Col. Hooker, AAG.

Captain Lyon's Report on the Pit River Expedition

Head Quarters, Pit River Expedition[67]
Camp near Lassen's Ranch, Cal.
August 1, 1850
To Maj. E. R. S. Canby
Assist. Adjt. Genl. 10th Mil. Dept.

Sir:

I respectfully submit the following brief account of the operations of this command since my report to you from Nicolaus reserving a detailed report to be accompanied with a map for a more convenient opportunity.

Having secured the transportation of the necessary supplies for the movement upon the head waters of Pit River, I left Nicolaus on the 8th June and marching four days arrived at this place.

[66]Peter Lassen's ranch ran along the east bank of the Sacramento River, where Deer Creek enters it, some twenty miles northwest of Chico, California.

[67]Lyon to Canby, August 1, 1850, NARG 393, LR, DP.

Up to this time I was still without an interpreter to communicate with the Indian tribes I might meet, and the means of ascertaining the Tribe or Tribes guilty of the murder of Capt. Warner, as also without a map or guide or other means of ascertaining the spot where he was murdered, except a brief note voluntarily given me by Lt. Williamson (as I passed Benicia) a copy of which will be forwarded on my accidental assignment to duty with this command, and my arrival soon after at Benicia, aware that the movement upon Clear Lake was already nearly a month behind the time contemplated in instructions of March 9th, I felt anxious that (so far as depended upon me) this delay should be no longer protracted and that the impatience that might naturally arise on the part of the authorities directing the movement might be removed and it was therefore with pleasure I received the announcement from Maj. Seawell that everything was in readiness for the movement, and I accordingly started the next day supposing myself possessed of all the means of intelligence it was possible to obtain, in the two months time from the date of the orders originating the expedition and its movement from Benicia.[68]

Upon closing the operation on Clear Lake, however, and contemplating those about to be undertaken to Pit River, I found myself sadly deficient in means to execute them so as to realize the objects contemplated in the instructions aforesaid of March 9th not having received a word or intimation of intelligence to govern me from Maj. Seawell.

I accordingly hastened (by express) to Benicia, my letter of May 22nd setting forth what I deemed necessary a copy of this letter was sent to Genl Smith, with a view to obtain such assistance as might be in his power to confer, accompanied with a request for one of the maps of Capt. Warner's route, a request made because Lt. Williamson told me there were in the possession of the Genl two on tracing paper besides the original. While at Nicolaus and on my way to this place, I made every effort within my power to remove the embarkments [embarrassments?] under which I was laboring, and mentioned in my letter of May 22nd without success, but received constant assurances that everything desired could probably be obtained at Lassens. On my arrival here

[68]While Lyon had wanted to acquire a copy of Williamson's map, he felt it necessary to launch the expedition before doing so.

however to my serious disappointment, I was informed by Mr. Lassen and others, whose long residence in the country entitled their statements to belief, that no person in the country could be found capable of speaking the language of the Indians (called "diggers") who live in the vicinity of where Capt Warner was killed, that no means were known nor could be suggested by which the Indians guilty of the murder could be identified, inasmuch as Capt. Warner's party, could not be distinguished from the many other small parties that were attacked in the vicinity of that place, and numbers of them murdered not only last season, but for the last four years that emigration has passed along this portion of the country, and that these Indians whoever they might be, were no more guilty enemies of the Whites, than all the Indians to be found upon Pit River and its tributaries, from near its mouth to past its head waters, and beyond the Sierra Nevadas to Salt Lake valley as all for several years have carried outrages against the Whites, to their utmost extent, and this last intelligence was confirmed by conversation with many emigrants of last year and so fully confirmed by our observation of several dead bodies along the trail of emigrants who were passing but a few days before—6 of a party of 7 having been killed.[69]

An Indian (half breed French) by the name of Inez, was mentioned as likely to be useful to me, not only as a guide, from having passed several times over this country, but possibly as an interpreter, as he could speak the "Snake" language, a language spoken from 50 to 150 miles north of "Goose Lake" Valley and through the medium of which, with more or less difficulty a communication with the Indians in view might possibly be opened.[70]

[69]Lyon is actually referring to two trails that converged on Lassen's ranch, one coming in from Salt Lake that entered California just south of Goose Lake (called Lassen's Trail) through Cedar Pass and then turned southeasterly through the mountains to the ranch. The Williamson party had met emigrants on this trail the year before. The second came down from Oregon, passing through to the headwaters of the Sacramento River and then to the ranch. Both were dangerous in these early years. It is interesting to note that a man like Lassen, who had been in California for nearly a decade, had so little knowledge of the Indians living northeast of him, near Goose Lake, or even of the geography that came to be known as "Lassen's Trail."

[70]The "Snake" Indians, as emigrants called them, were actually Paiutes and Shoshones from Idaho and eastern Oregon. Having an interpreter who spoke their language likely would have been of little use in communicating with Modocs or Klamath Indians who often entered the regions around Goose Lake.

Inez was sent for with great industry, causing me some delay without success. As a last alternative I procured an Indian who could speak the language of a tribe a short distance in the mountains, with a hope to procure one of this tribe to form a connecting link with the next tribe, from which a link might be procured in like manner, till a succession of interpreters might be secured to answer the purpose; this attempt however failed from not being able to secure a member of the first tribe. I therefore found myself on my way and already in the mountains, without a guide or interpreter, or means to discriminate between the murderers of Capt Warner and other Indians—On leaving this place (June 16th) a note overtook me from Mr. Avaline at Sacramento City stating that he wrote in consequence of the enquiries that he heard had been made in regard to the remains of Capt Warner, and that he was present when Capt Warner was killed and could guide me to the spot, but would wait till he could hear from us and learn whether the terms in which he would be employed would be satisfactory.[71] Tho the desire to show respect for the remains of a much lamented brother officer, and the note from Genl Smith requesting particular attention to be given to this matter, would awaken a keen desire and authorize me to make every possible effort to recover them, I did not consider this the main object of the expedition; "the infliction of such punishment upon the guilty tribes as will effectually deter others."[72]

No attention was paid to this note of Mr Avaline, not only because of the aggravated delay that would be occasioned thereby, but for the reasons above given in which I was confirmed from the fact that the authority directing this movement and to whom the existence and probable availability of this and other individuals connected with the operations of Capt Warner at the time of his death, were likely to have been known, had not deemed the matter of sufficient importance, to express or intimate any intelligence upon this subject, during the three months time from the organization of this command to the time of its leaving Nicolaus for the Head Waters of Pit River. Under those

[71]Mr. Avaline was a civilian who had accompanied Warner into the Goose Lake region.

[72]The need to find Warner's body increased after his brother turned up in San Francisco with the hope of bringing the remains back east on his return. Lyon obviously had not been pressed on this subject earlier, seeing his mission as one of "chastisement," much like with the Pomo Indians.

circumstances I determined to make the movement directed, and to repell [*sic*] the hostilities of which Emigrants and others so much complained should my experience confirm their statement, and to chastise these Indians so as to drive them off the trail, and prevent their infesting it and committing outrages upon the emigration of this season, and to deal seriously with the hostile bands in the vicinity where Capt Warner was murdered. The accounts of hostile Indians were fully realized and on my return the country on both sides of the trail was scoured 12 Indians were killed and others wounded and one horse and a mule (stolen from emigrants) taken from them, two men of my command were severely wounded, one of whom died the 2nd day after; particulars of these operations will hereafter be given.[73]

Guided by the notes of Lt. Williamson I sought for the place at which Capt. Warner was murdered and so far as I can now learn from Mr. Barling (one of Capt. Warner's party) I passed near if not over the precise spot, but found no trace of him, nor of the operations of his party, or that of Indians; and during the three days movements that this search was continued no signs of probable Indian presence within a long time and apparently the last five years could be discovered, except the appearance at a single point of one Indian having passed about three weeks previous, having lain down beside a stream where he had eaten some small animal.[74]

On returning to this point on the 30th ult., I found increased importance attached to the recovery of the remains of the late Capt. Warner,

[73]Lyon did not make a second report, and it can only be assumed that the Indians killed were part of small group of so-called Pit River Indians, known today as the Achomawi people. They were often at war with the Modoc and Klamath to the north and west, and these latter tribes frequently took Achomawi people captive, bartering them at the great slave market located at The Dalles, on the Columbia River. By reading Williamson's report as well as Lyon's, it seems obvious that slave raiding and disease had decimated these people to the point where populations were very small. Williamson never even mentions them by name, Lyon is unaware of their identity, and Lassen knows practically nothing about them. See D. L. Olmstead and Omer C. Stewart, "Achomawi," in Heizer, *California*, 225–35. Kroeber's commentary is suggestive of rapid extinction. "No vocabularies were recorded," he noted, "the tribes being numerically insignificant, and in one case on the verge of extinction when the white man came to northern California." See Kroeber, *Handbook of California Indians*, 279.

[74]Lyon is here referring to the upper Pit River and regions around Goose Lake, where he found no evidence of Indian occupation.

from the orders from Govt on the subject, and the anxiety of the General Comdy of the Division, apparent not only from his communication, but his repeated (but ineffectual) efforts to push forward to me these communications and the two men Avaline and Barling, who were with Capt. Warner at the time of his death. With a view to realization of this object, with as little delay as possible I have directed a detail of 24 men to be retained at this place, to await the further orders of the Division Commander, to whom an express was dispatched by Lt. Stoneman early on the morning of yesterday, the remainder are dispatched to the respective posts to which they belong. Orders marked 1 & 2 herewith enclosed give the necessary directions. I shall remain here awaiting the return of Lt. Stoneman, and anticipating orders to return. I shall be entirely prepared therefore and will probably move in about 4 or 5 days, expecting to be absent from here about 30 days. Barling is here and Avaline probably available, he is sent for.

Enclosed herewith I transmit monthly reports of this command for the month of June & July and a list of Deserters during the first 15 days of June inclusive.

Very Respectfully
Your Most Obt Servt.
(signed) N. Lyon
Bvt, Capt. 2nd Infty
Comdy Expedition

P.S. the description of the deserters is not given because the description books of the companies are not along with the command.

(signed) N. L. Bvt. Capt. 2nd Infty.

P.S. 2d (Postscript 2d recommends the establishment of a post at the Head of Sacramento Valley giving some reason therefore—

(signed) N. L. Bvt. Capt. 2d Infty

Erasmus D. Keyes's Expedition into Southeastern California May–June 1851

While Captain Lyon was reconnoitering the Pit River valley in the northeast, an opportunity to explore far southeastern California fell to Captain Erasmus D. Keyes. He joined members of a peace and treaty commission in 1851 to explore and map the regions south of the bend of the San Joaquin River. The three men of the commission, George M. Barbour, Oliver M. Wozencraft, and Redick McKee, selected different parts of California for their work. Keyes joined Barbour with an escort, given the belief no doubt emanating from Derby that the Natives of the Kern River valley might prove hostile. While Captain Keyes's journal of this expedition has survived, he left only a rough sketch of the trip, completed by one of his young lieutenants, which did not improve on Derby's earlier efforts.

"Report of Captain E [Erasmus] D. Keyes, Commanding Escort for Col. Barbour"[75]
Camp Magruder, near
Los Angeles, Cal.
June 17, 1851
Captain F. Steele,[76]
A.A.A.G. 10 Mil. Dept.

Sir:

I have the honor to report that I arrived here with my command, escorting the Indian Commissioner, the 16th instant. Col. Barbour has this day dispensed with the further services of the Escort, and I shall as soon as necessary repair our completed [complement] of wagons, harnesses, &c. and we have received a few additional supplies from

[75]Keyes to Steele, June 17, 1851, NARG 393, LR, DP.

[76]Captain Frederick Steele was post adjutant at the time of Keyes's expedition. He later served with General Lyon, fighting at Wilson's Creek in Missouri for the Union. He rose to the rank of major general during the Civil War, commanding troops mostly in the West, and after the war returned to California.

Captain Lovell's Post at Chino, [and will] commence the march up the coast to Monterey, San Francisco, & Benicia.[77]

Since leaving Rio de los [Santos] Reyes, the 27th May, I have made no report of my operations, as our quartermaster's force is small, and the expense of sending dispatches would have been considerable.[78] This report will therefore embrace what I have to communicate concerning the march &c. from that point to this.

On the 30th day of May, 1851, Col. Barbour concluded a treaty with eight tribes of Indians at Camp Keyes on the Cahwia River.[79] These tribes or bands are generally called the Four Creek Indians and are more disposed to hostility than any we have seen. Their numbers were estimated at about 1240 individuals, or souls, if such monstrous creatures have souls. In the treaty, the commissioners set apart for the exclusive use of the Indians reserving the right of way and to establish military posts, two strips of land on the east and west of former reservations made at the Treaties of Camps Barbour and Belt. On the west, therefore, the Indian reservation is bounded by a line running from the extremity of Tulare or Tache Lake, near the mouth of King's River, to the mouth of the slough at the great bend of the San Joaquin river, and on the east, by a line running from the two ponds or small lakes at the head of Cahwia river to the nearest point of Kings river. After examination, I would recommend the establishment of a military post at or near the bridge over the Cahwia river where the treaty was made and on the right bank of the river, or at some point on the hills one mile further up the same river. Major Miller's camp was directly opposite on the left bank, and it is thought by some that it had a less healthy site than either of the two I propose.[80]

77Captain Christopher S. Lovell received orders to establish a military post east of Los Angeles, his efforts based on recommendations from Major Ord. The post was near present-day Pomona and Chino, and it was initially seen as a check on the Cahuilla Indians, whom Lovell later visited and wrote about. See Robert W. Frazer, ed., "Lovell's Report on the Cahuilla Indians, 1854," *San Diego Historical Quarterly* 22 (Winter 1976).

78Río de los Reyes is an early appellation for King's River.

79The Cahwia River was in the Four Creeks region just south of King's River. It ran from what is today Lake Kaweah westward, into the Tulares Lake region.

80There is some controversy regarding the location of these two sites, Camp Barbour and Fort Miller, some historians suggesting that they were in the exact same locale. Keyes seems to settle the issue by placing Camp Barbour on the north bank of the San Joaquin River and Camp Miller on the south. The military records in NARG 393, LR, DP, suggest that Camp Miller was thereafter moved, which helps explain the confusion.

The land on the banks of the Cahwia is rich, and the wood and vegetation much more abundant than I have elsewhere seen on our route. Timber of an inferior quality for building may be had and the water is unfailing. The distance of the crossing of the Cahwia from Stockton, from which point supplies should for the present be brought, I should think is a little more than two hundred miles—the rivers in the dry season are excellent, but in wet weather transportation over them would be very difficult. The Four Creek country is reputed to be somewhat unhealthy from June to November, and appearances made such a supposition quite probable.[81] For that reason, good quarter should be constructed for officers and men. I would further respectfully recommend that the Post be garrisoned by at least one full company of mounted men, detachments from which should be kept continuously in motion about the Reserve. Foot soldiers would be nearly worthless, and horses or mules may be kept cheaply on the Cahwia.

On the 3d of June, the Commissioner concluded a treaty with four tribes or bands at Paint Creek, about 40 miles south from the Cahwia, and a little more than 50 miles northerly from the Texon [Tejon] Pass.[82] The lands reserved for the four tribes are in two distinct reservations; one lying between Tulare & Buena Vista Lakes, and the other being bounded south by Paint Creek—west by the Emigrant Road, north by Tule River, and east by the Sierra Nevada Mountains. The estimated number of the four tribes is 1660, less disposed to be hostile than the Four Creeks Indians. The country reserved for them is generally poor, as far as I could see and learn, and I concur in opinion with the Commissioner that no military Posts are required among them, as they will be under the influence of the Posts on the Cahwia and at the Texon [Tejon].

The next and last Treaty concluded by Col. Barbour was at the westerly entrance of the Texon Pass [usually Tejon Pass] with twelve bands or tribes, embracing 597 individuals. The land embraced in this reservation is bounded by the Kern river, starting from its first fork near

[81]The reference here is to malaria, which had become epidemic in the years previous to the treaty signing. The epidemic accounts for the small number of Indians reported in the region by Keyes.

[82]This treaty was negotiated at Tule River, which has since disappeared due to irrigation. The Tule River Indian Reservation is roughly in the region where the river ran.

the Tar Spring to Carises lake, thence by a line to Buena Vista Lake, thence from the most westerly point of said Lake to the nearest point of the Coast Range, thence along the base of said range to the westerly entrance of the Texon Pass—thence in a straight line to the place of beginning.[83] Some of this land is quite good, but that on and in the neighborhood of Kern River is little better than a desert. I recommend that a post be established at or near what is known as French's Rancho, where the wood, water, & grass are always in great abundance, and where the climate is quite healthy. This post might best be supplied from San Pedro, about 90 miles distant, the road from San Pedro, about 90 miles distant.[84] The road over the mountains is very steep and rugged in many places, but with a modest outlay, it could be rendered very much better. Supplies must be sent over before the rainy season sets in, and the loads for six mule wagons should not exceed 1200 lbs. each.

As soon as the map and memoir of the country over which we have passed shall be completed by Lieut. Hamilton, 3d Inf., the topographical officer to the command, I will write more fully in relation to various tribes of Indians with whom treaties have been made.[85]

I shall leave here with the command and entire train before the 25th inst.

Very Respectfully
Your Obed. Servt.
E. D. Keyes
Captain, 3d Inf.
Comd Escort

[83]Both Carises and Buena Vista Lakes have been drained of water today. Carises Lake was likely just east of the upper forks of the Kern, near what is today a series of hot springs. Buena Vista Lake was roughly twenty miles southwest of Bakersfield, California.

[84]It is impossible to identify the location that Keyes is referring to, somewhere south of Kern River. The Campo de los Franceses, a land grant to one William Gulnac in 1844, is in San Joaquin County well to the north. It is the only grant that might translate "French rancho." See Cowan, *Ranchos of California*, 22.

[85]The map referred to is not with the journal in NARG 393, LR, DP.

Major Wessells's Exploration North of Sonoma August–November 1851

While Captain Keyes was reconnoitering the regions to the south of the San Joaquin River, another expedition was headed into regions north of Sonoma. There was growing activity and interest in this region, as it was practically unexplored. Ports for embarkation along the ocean were far north of San Francisco, the most frequently used being Humboldt Bay. To the east of the bay was a vast expanse of mountains and river valleys—the Eel, Klamath, Trinity, Salmon, and Scott Rivers, in particular.

It was generally believed that substantial Native populations lived in these valleys, and the army wanted information on their numbers and locations. The job was assigned to Bvt. Major H. W. Wessells, who provided escort for Commissioner Redick McKee, who intended to negotiate with the various tribes in the region.

"Report of Bvt Maj. H. W. Wessells, Comdy Escort of Col. McKee on tour of Klamath, Trinity, Humboldt, etc"[86]
Post Near Benicia, Cal
November 14, 1851
Bvt Lt. Col. J. Hooker
Asst Adjt. General
Pacific Division

Sir:

Agreeable to instructions from Headquarters Pacific Division, of 31th July last, I have the honor to report that in obedience to Division Orders of that date, I proceeded to Sonoma, and on 8th of August assumed command of a Detachment of thirty-six privates 1st U.S. Dragoons, destined to as an escort for Redick McKee, an agent of the Indian Department, who was desirous of visiting such Indian tribes, as inhabit the waters of Clear Lake, Russian and Eel Rivers, Humboldt Bay and

[86]Henry Walton Wessells graduated from West Point in 1833, serving in the army of the frontier thereafter. He later reached the rank of brigadier general during the Civil War but was captured, exchanged, and later organized the draft in New York. See Wessells to Hooker, November 14, 1851, NARG 393, LR, DP.

if the season would permit, the Indians of Klamath and Trinity Rivers. The object of the visit was to make treaties of peace, reservations of land, to issue provisions, clothing and other presents for their temporary use, also to make provisions for their welfare and comfort.

The agent having reported himself ready, the command marched from Sonoma on the morning of August 11th, with a well appointed train of pack mules, three wagons, and provisions for thirty days. The agent was also provided from the Qr. Masters Department with a separate train of mules, for the transportation of his private stores, and Indian goods—Asst Surgeon Griffin U.S. Army, accompanied the Expedition.[87]

The command proceeded up the valley of Sonoma, and through those of Williams and Santa Rosa, into the Valley of Russian river, each of which is gradually merged into the one succeeding it, affording an excellent wagon road, through a country of great beauty and fertility, fast filling up with an enterprising and industrious class of citizens.[88] Russian river was crossed at Fitch's rancho, and on the 14th the command remained in camp on the river at the Soda Spring, about 50 m. from Sonoma, in order to meet Indian runners, who had been previously sent forward to notify the Clear Lake tribes of our approach—a large herd of cattle intended for issue to the Indians here joined the expedition, prepared to follow it across the mountains.[89]

The 15th and 16th days of August, were consumed in crossing the hills which for the lower cañon of Russian river, a distance of 22 miles by what is called the wagon road, but which for business purposes is totally impracticable for other transportation than pack mules. In order to meet the tribes about Clear Lake a temporary depot was established at the Rancho of Fernando Feliz, and on the 17th a portion of the escort accompanied the agent to that place, a distance of 15 miles, it being separated from the waters of the Russian River by a

87It should be noted that McKee carried along presents for distribution to the Indians and food for feasting. This made it very difficult for reticent bands to resist coming to the anticipated negotiations.

88The Santa Rosa valley was a mere twenty miles northwest of Sonoma and the Williams valley just beyond it.

89Henry Fitch received a grant for Rancho Sotoyome in 1844. The ranch was located near Healdsburg, California, and was managed by Moses Carson in 1851. See note 48.

ridge of considerable height and on its eastern declivity almost inaccessible. Having formed a camp in the valley, about two miles from Kelsey's Rancho [Kelseyville, California] several conferences were there held with representatives of Clear Lake Indians, or Lu-pi-yu-ma (Stone) numbering according to their own account, about 700 souls, though they have been generally represented by the Whites, as exceeding 1500 and even 2000.[90] These tribes, though occupying one of the most beautiful and fertile valley's in the state, abounding with game, and susceptible of cultivation throughout its whole extent, and which borders upon a lake 30 miles in extent, well stocked with fish, present an appearance of squalid misery, calculated to excite pity and commiseration, rather than a single feeling of hostility.[91] The obligations of a treaty were explained to them, as far as their stupid ignorance would permit, and on the 20th a paper was executed and duly signed by the principal chiefs, defining the limits of a reserve on which they were to reside as their own possession, stipulating also, for the removal to it, of all the tribes on Russian and headwaters of Eel Rivers.[92] The treaty further provided for a resident agent, to whom they were to refer all grievances, and an annual distribution of beef and flour, with working oxen, and certain implements of husbandry. These Indians though of indolent and brutal habits, are extremely docile and by judicious management, out of the hands of speculators, could be made useful as laborers particularly in the care of stock.

On the 20th of August the detachment joined the main body and on the day following, a council was held with some 200 Indians representing a portion of the Russian River tribes; on the 22d a treaty similar in

[90]Both the ranches of Fernando Feliz and Andrew Kelsey were in the valley between Russian River and Clear Lake, both visible on Lyon's map of 1850. The Habinapo Indians, whom Lyon assaulted in 1850, were called by the nearby Miwok the Lupuyuma, which is the word that Wessells applies to them. A literal translation of Habinapo is "Stone People." See Kroeber, *Handbook of California Indians*, 232.

[91]It should be noted that part of the reason for the difficulties of these Indians resulted from the attack the year before by Captain Lyon, when large amounts of stored food and housing had been destroyed. Wessells seems oblivious to Lyon's attack.

[92]It is unclear from Wessells's account whether the Pomo Indians on the headwaters of the Eel River were at the negotiation. From the population figure, a mere 700 people, it would appear that they were not party to the agreement.

character to the former was signed by the chiefs present, a part of the bands only agreeing to remove to the reservation on Clear Lake, though it is very doubtful if they understood the full import of the condition.

As well as could be ascertained, these tribes number from 1000 to 1100 souls similar in appearance and habits to those of Clear Lake but perhaps less docile and more stupidly ignorant; many of them have been for years employed or have been residents on the different ranches of the valley, and seem to have become identified with the estates, though producing little in proportion to the numbers so employed.[93]

The tribes thus far met with seem to possess a common dialect, though in some few instances it was found necessary to employ a second interpreter. Clothing, though eagerly worn, seems to be a matter of little importance to the males, the only article manufactured by themselves being a short mantle called Steet, made of the skins of hares, possessing nothing of sufficient value to barter they obtain little except occasional presents from the whites, or from the rancheros who employ them.[94] The march was resumed on the 23d and the wagons having with great difficulty been brought thus far were sent back to the depot at Benicia, the route hence for about 20 miles lay along the right bank of Russian river, and its south Fork, thence across the dividing ridge between it and Eel River, a rough and broken country, though letting down gradually, into the valley of Be-tum-ki, which contains the headwaters of the latter stream. Our camp formed at the lower end of the valley, was visited by Indians in considerable numbers, and two days were expended in futile attempts to explain and enter into a treaty of peace and friendship with their several chiefs; they could in nowise comprehend its obligations, and positively refused to leave their native valley—five tribes numbering about 400 souls were represented

---

[93]Wessells's estimate is similar to Commissioner McKee's, who put the entire population of the Russian River Pomos at about 1,200 souls. See McKee Report, November 15, 1851, NARG 75, LR, California Superintendency. Kroeber has a broader, but not necessarily higher, estimate. He saw the McKee report but put the total population of the Pomos and Wappos from Clear Lake, the Pomos from Russian River, and a small number who still lived west of Russian River on the coast, at 2,500 Indians. See Kroeber, *Handbook of California Indians*, 237.

[94]Wessells seems very certain about the mantle worn by men being made from the skin of rabbits. Kroeber learned that it was more common for redwood bark to be employed, although rabbit skin seems more feasible. See Kroeber, *Handbook of California Indians*, 240.

in council, they seem to have had but little intercourse with whites, were evidently suspicious of our intentions, and declined bringing in their women & children.

This valley, about four miles in length, well wooded and watered, communicates with another of greater extent through a difficult cañon, which it was necessary to turn, and although the distance is less than 20 miles, three days were consumed in effecting a passage, the route being difficult and our guide inexperienced. This second valley called by the natives, Be-tum-ti-ki-ah was reached on the 30th and our camp was freely visited by its inhabitants, and with them, another fruitless effort was made to form a treaty. They could not be made to understand the obligations of a contract, and were very suspicious of our designs. These tribes probably number 500 souls, have had little or no intercourse with the whites and in every respect appear to advantage when compared with those living among the settlements, their women and children as at Be-tum-ki, were not seen, nor could they be induced to bring them in.[95] The men are in a state of nature, unusual I believe in any of the most savage tribes of our continent, scarcely an article of clothing being seen, the moccasin and breechcloth are unknown, and a single thread fastened close around the waist, is all the clothing they wear, though even in summer the nights are cold. They are quick and active in all their movements, well armed with bows and arrows, but though living in a country abounding with game, seem to have little success in hunting, or at least make no use of skins. Small parties

[95]Kroeber translates Betumki as simply the Pomo word for "Large Valley," and Be-tum-ti-ki-ah was likely a second rendition of "Large Valley," with some added reference. Several Yuki rancherias were located in the headwaters of the Eel River, where the party negotiates the first set of treaties. While a number of 400 people was likely a guess on the part of Wessells, the population of these Yuki Indians, who inhabited the headwaters, was obviously not large. The references to "scourges" entering the region as early as 1848 by Major Hardie suggests disease as a factor in depopulation. Wessells's figure of 500 for the more central valley of the Eel River is also small, and the lack of populations of any size farther down the river toward the coast suggests a total population for the Yukis of perhaps a few more than a thousand people. Some historians, however, have suggested extremely large populations, putting the number of Yukis at 5,000 to even 20,000 people. There is absolutely no evidence to support this claim. See, for comparison, Ben Madley, "California's Yuki Indians: Defining Genocide in Native American History," 303–42. As part of the Pomo group, the Yuki language—but not population—is discussed in McLendon and Oswalt, "Pomo: Introduction," 282.

passing through their country should do so rapidly, and be constantly on the alert, for they are a thievish race, and with sufficient temptation, may proceed to extremities, provided success is very apparent.

In crossing the range, between the two valleys a species of chesnut [chestnut] tree was met with, bearing burs in clusters, each containing a nut, similar to the chincopin [chinquapin] of the middle states.

Not being able to effect an arrangement with these Indians, the march was resumed September 1st, and leaving the valley, entered upon the coast range; the route hence being a continued succession of mountains, though fortunately affording an abundant supply of excellent grass and water. A few Indians were seen at distant points, but either from fear or suspicion, could not be induced to visit our camps.

This whole region abounds with game, such as bears, elk, and deer, and in the fir forests, the pinated [pinnated] grouse is frequently seen—a wild cherry, of the astringent kind, grows in these mountains.

In consequence of rain, and to refresh the animals, we encamped on S. Fork of Eel River during the 4th of September (at this place a considerable stream, even at low water, being some 30 yards in width) and on the following day crossed a mountainous bend of the same stream, encamping on it again in a small valley, some 20 miles below. Here a few Indians came in but having no interpreter, they were dismissed with presents, and without being able to obtain from them any information as to their numbers, conditions, etc.—having halted at this place another day on account of rain, the march was continued two days down the bed of the river, and on the 7th we encamped on the main stream, four miles below the junction of its forks.[96] This route in time of high water would be attended with great difficulty, the mountain spurs (almost impassable in many instances) jutting out so abruptly into the stream that to turn them it becomes necessary to cross it. At this season, however, the river being fordable at almost any point, we experienced little difficulty except from the stony character

[96]The party is in the heart of the Round valley, just south of Dos Rios, California, on the main Eel River. Again, the few numbers of Indians seen and reported is indicative of small populations, especially given the fact that the party spent nearly a week in this location, hunting and resting. Also of note is the fact that while they found a few canoes used for fishing, there were no villages, or even remains of villages.

of its bed, and the trouble of frequent crossing. Indians were seen in small numbers, but there were indications of its being much frequented for the purpose of fishing, a few canoes were seen at the junction.[97] No portion of this region is suitable for agricultural purposes, being mountainous and heavily timbered with forests of redwoods. The trees are of gigantic size, one of which lying on the surface of the ground contained a diameter higher than my head, though mounted on a horse of large size and another measured 325 feet in length.

The route still lay along the river bed, as far as Van Densens [Van Duzen's] fork a large tributary coming in from the north and thence down its right bank to a camp ten miles from the ocean and about the same distance from the town of Humboldt [the south end of Humboldt Bay near Eureka] the lands adjacent to Van Densen's Fork [Van Duzen's Fork], and the main river below it are well calculated for farming purposes well timbered with Oak and Redwood, with abundance of excellent water, and are already occupied by a considerable number of thriving settlers. The mouth of Eel River is not accessible for seagoing vessels.

Another fruitless attempt was made to form a treaty with the tribes inhabiting the low waters of this river, but our means of communication were so imperfect, and the different bands evinced so much jealousy of each other, that nothing was affected, though five days were consumed in endeavoring to bring them to terms. These tribes numbering about 300 souls, have generally been on friendly relations with the whites since the settlement of that portion of the country, and seem to be under good subjection, few complaints being heard for aggressions of either side.[98] They are however as low on the scale of humanity as is possible to conceive, of brutish habits, hideous repulsive features, and loathsome from disease.[99]

---

[97]This referencc is to the junction of the main branch of the Eel River and the Middle Fork, which entered it from the east.

[98]Once again, the population numbers—a mere 300 souls—suggest massive depopulation from disease. The Indians that the party negotiated with were likely Wiyots, who inhabited the lower Eel River and Humboldt Bay. See Albert B. Elsasser, "Wiyot," in Heizer, ed., *California*, 155–63.

[99]Wessells does not identify the disease, but later evidence, especially from McKee's journal, strongly suggests malaria. The month of September was usually the time of its eruption.

To this point, which should be reached in fifteen marches, the distance from Sonoma was computed at 240 miles, and little in convenience will be experienced in crossing over except from the altitude and steepness of the mountains. The climate throughout is cool and agreeable, grass and water abundant and of excellent quality, and annoyance from insects almost unknown. Though well stocked with almost every variety of game, the natives seem to depend entirely upon acorns, grasshoppers, roots, and grass seeds for their sustenance, except those near the coast, who subsist chiefly upon fish and clams.

The march was resumed on the 15th, via the town of Humboldt, a small settlement on a tongue of land near the lower extremity of the bay, and opposite its entrance; thence skirting the bay across the Eel River, which at low tide is fordable near its mouth to Bucksport, and from that place to Union town, over a difficult trail across the mountain spurs, and an extensive marsh, interrupted by tide waters from the bay, the distance being from Humboldt about 20 miles. Arrived at Uniontown, the march was again delayed to enable the agent to transport his goods from Port Trinidad, some 18 miles up the coast, and the point to which they had been shipped.[100] The animals were turned out to graze, but though grass was plenty, they were so harassed and worried by mosquitoes, as to fall off both in flesh and spirit during our stay here.[101] On the 23d the march was again resumed in a northerly direction, crossing a succession of mountains, separating from each other, Mad River, Redwood, Pine Creek and the Klamath, and halted at Bloody Camp, an elevated position, two miles from the junction of the Trinity and Klamath, and the nearest point to that place affording grass. It was here discovered that a considerable quantity of Indian goods were yet to be transported from Port Trinidad, and accordingly

[100]Bucksport and Union Town both disappeared in the nineteenth century, primarily because their location was on the bay, which was not capable of handling larger ships. Thus Trinidad became the main shipping point. Bucksport was built in 1849 on the former location of a Wiyot Indian village, and the army constructed Fort Humboldt above it on the bluff. See Hoover, Rensch, and Rensch, *Historic Spots,* 245.

[101]Wessells seems to contradict himself here, noting the vicious mosquitoes but emphasizing the lack of them somewhat earlier. The mosquitoes spread malaria in the region, which carried well up the various river valleys, as the party would see when on the upper Trinity River.

a train of mules was dispatched on the 28th and to communicate more readily with the Indians, moved camp on the following day to the mouth of Trinity, leaving the horses and pack train with a guard on the mountain.[102]

The junction of these two rivers is at the bottom of an immense chasm, almost inaccessible in its approach from every direction, and but for the persevering search after gold deposits would scarcely have been trodden by the foot of white men. It occupies a central and commanding position with regard to numerous tribes on both rivers above, and on the Klamath itself below the junction. As a military point it would exercise a controlling influence over the tribes above named, as well as the hostile bands infesting the waters of Redwood. Unfortunately, however, it possess no resources for the establishment of a permanent post, the character of the ground being unsuitable for gardening, or even building purposes. The expense of supplying a garrison would be enormous, and it may be doubted if there is sufficient interest in that quarter at present to justify such an expenditure. In view of quieting all difficulties and animosities between these Indians and such miners and packers as occasionally pass through their grounds, I would respectfully suggest that a small force well mounted on mules, forming a moveable camp be sent through that country when the season will permit, authorized to issue a limited quantity of provisions and suitable for halting at central points long enough to familiarize those savages to the presence of troops, who, by their peaceful conduct will inspire confidence as friends, and show by their arms and discipline, that they are able not only to protect themselves, but to resent all injuries and impositions. A temporary depot, established at Bucksport or Uniontown, could supply the camp by means of pack train; the former town is on the eastern shore of the bay, and at all times accessible by water, whilst the latter is situated at its northern extremity, a day's march nearer the Klamath, but inaccessible for anything but small launches. I would also venture to recommend the same course, with respect to those Indians living on the upper waters

[102]The party is at the junction of the Klamath and Trinity Rivers, having moved overland from Humboldt Bay to reach that location. It held a concentration of small Indian rancherias.

of Eel River, in order to accustom them to the presence of soldiers, to teach them the obligations of a bargain or contract, and if possible remove their fears and suspicions.

The Klamath and Trinity Indians, with a few exceptions, came freely into camp bringing their women and children, and exhibiting an appearance of open, cheerful frankness, entirely different from that of any tribes heretofore met with. They are generally of a light copper color, straight and well formed, and apparently of active habits, each warrior has a bow and arrows, armed with flint or iron heads, besides a long dagger, or kind of "Couteau de chasse," of foreign manufacture, which it seems they have obtained from occasional intercourse with more northern tribes. Mantles of dressed deer skins are worn by both sexes, and in some instances, ornaments of feathers on the head—their dwellings are simple in construction, but far superior to those of any tribes in the upper part of California, an excavation some ten or twelve feet square being made in the ground, a house of boards placed upright along its sides is formed, with sloping roofs and entirely closed, except a circular opening in front which permits ingress, and egress to the occupants, a pavement of smooth stones is always laid before each dwelling. The bodies of deceased friends and relatives are enclosed in a rude box of boards and deposited in the ground in the immediate vicinity of their houses, the grave being ornamented with rows of baskets securely fastened and frequently enclosed by a rectangular paling, the upper end of each picket being decorated with the feathers of birds; ornaments formerly worn by the deceased are also suspended within the enclosure. Their fishing dams exhibit considerable skill and great perseverance, being formed by timbers laid horizontally, securely fastened together and strengthened by uprights; the whole extending across the river and necessarily possessing great solidity in order to resist so rapid a torrent as the Klamath.[103]

---

[103]Obviously, Wessells was somewhat taken by the industry of these Native Americans. They were represented by three distinct groups, the Chimarikos, near the junction, the Yuroks, above it on the Klamath River, and the Wintuns, above the Chimarikos on the upper Trinity River. A *couteau de chasse* is a French hunting knife, likely similar to those carried by French mixed-bloods who commonly came into California as part of fur-trading expeditions. See Shirley Silver, "Chimariko," 205–10, and Arnold R. Pilling, "Yurok," 137–54, both in Heizer, ed., *California*.

Salmon are taken in large quantities and in some seasons constitute their chief article of food. These fishing dams however, prove a constant source of trouble and complaint, the upper tribes being often dissatisfied because the salmon are obstructed in their passage by the dams below them. Their females are virtuous and as far as we could observe, during our stay among them, behaved with propriety; in fact, the whole people manifested much satisfaction in their intercourse with the command, it being the first detachment of soldiers they had ever seen. The dress of women is uniformly the same, consisting of a petticoat of dressed deer skin, fastened around the waist, and reaching nearly to the knees, with a thick fringe of the same material extending to the calf of the leg, the upper edge falling over and embroidered with threads of grass differently colored—to complete the costume an apron is worn, formed by strands of braided grass and ornamented with horizontal rows of small shells. A small saucer shaped basket of woven grass is worn on the head and all are tattooed with three lines of deep blue, from the corners of the mouth and centre of the under lip, reaching below the chin, these lines being narrow in childhood and widened as they advance in years. Our means of communicating with these tribes were so imperfect that it was impossible to gain an accurate knowledge of their customs, ceremonies, etc., or their notions as to their own origin, or a future state. I was also unable to ascertain their numbers, and it is extremely doubtful if any reliable information on that point can at present be obtained.[104]

After much delay arising from misunderstanding and jealousies between different tribes, a treaty of peace and friendship was formally signed on the 8th, by such chiefs as were willing to do so, and a reservation marked out, embracing the point where the two rivers join, and extending along the crests of the mountains for some miles above and below on each of them.[105] It is impossible at this time to find any one,

[104]While Wessells refused to venture a guess on population, McKee reported the entire population of the region both above and below the junction of the two rivers as not more than 1,500 Indians. McKee received this information from C. W. Durkee, who had lived in the region for some time and ran a ferry at the junction of the two rivers. Durkee knew the various Indian languages of the region and continued on with McKee as interpreter while the party moved farther up the Klamath and ultimately the Salmon River. See McKee Report, November 15, 1851, NARG 75, LR, California Superintendency.

[105]This location was similar to the Klamath River Indian Reservation established for these Indians in 1856.

sufficiently familiar with their language, to explain abstract ideas, or to teach them according to our notions, the obligations of a bargain or contract, and it is hardly probable that they had a very clear conception of the character of the instruments to which their marks are affixed; though I believe them to be sincere in their promises to remain good friends in future. Mr. Durkee, an enterprising adventurer from New England, has resided on these rivers for several months, and by good management has acquired considerable influence among the tribes near the junction, he has also some knowledge of their language, and in due time, may qualify himself as to render valuable service as an interpreter. There is also a chief called Morucas, at the head of the Trinity, whose fearless and determined character renders him feared & respected by the bands around, which controlling influence with his generally friendly deportment towards the whites, may prove valuable in future intercourse with that people—the distance from this point to Uniontown was estimated at 45 miles.

On the 9th of October the command crossed the Klamath by means of a good ferry established just below the junction and continued on up its right bank almost due north.[106] On the 11th encamped nearly opposite the mouth of Salmon, a considerable stream coming in from the east—this route is both difficult and dangerous; several horses and mules were precipitated from the side of a mountain several hundred feet below, and some of them crushed to death by the fall. At this point a treaty was formally signed by chiefs representing several tribes in that vicinity, similar in appearances to those below, but differing somewhat in language, which rendered communication with them still more difficult and imperfect.[107] These Indians were admitted into the former treaty by an addendum and entitled to all the privileges therein conferred.

The Klamath River throughout all its course is a rapid torrent, sweeping the bases of an almost endless succession of precipitous mountains,

[106]Durkee became invaluable to McKee as an interpreter and source of information. Wessells seems to trust him as well, noting his enterprising personality. The ferry referred to by Wessells is the one established by Durkee.

[107]The Natives at the mouth of the Salmon were Karok Indians, closely related to the Yuroks of the lower Klamath. See William Bright, "Karok," in Heizer, ed., *California*, 180–89.

which rarely recede from it a sufficient distance to allow a pathway between them and its margins; hence the trail must either pass over their summits, or wind along their nearly perpendicular sides. The same description will also apply to its tributaries on either side.

All the bars are more or less rich in gold but the great expense attending the transportation of supplies will for a long time render them of little value to the miner.

The march was resumed on the 12th of October continuing along the river and the trail so difficult that it was necessary every day to send in advance a strong pioneer party to render it passable for the horses, which were now found to be entirely useless in a country of such rugged character and affording so little nourishment. They also suffered much and caused considerable delay in crossing a succession of low rocky points, whilst the pack mules were over them without much difficulty or detention. On the 19th we encamped two miles below the mouth of Scott River and 95 above Trinity, having the day previous crossed to the left bank of the Klamath, the latter stream being nowhere fordable below the mouth of Scott's—though readily passed by means of canoes. A few Indians were seen along the river, peaceable in disposition, but an inferior race to the tribes lower down, small parties of miners were also quietly at work on the different bars, and there seemed to be no signs of hostility on either side. At Scott's Bar four miles above the mouth of that river, a considerable mining town has sprung up and the efforts of its inhabitants have been attended with great success though its population has been recently much diminished by the excitement consequent to the discovery of quartz veins in various parts of that country.[108]

After passing the bar, the route continues up Scotts River two or three miles and thence over a mountainous bend, with a valley of the same name its lower extremity being about 15 miles from the Klamath. We encamped here on the 20th.

The agent, desiring to return to San Francisco by the way of Port Trinidad, a verbal arrangement was made between us before leaving

[108]Scott's Bar survived and is today located four miles south of the junction of the Scott and Klamath Rivers.

the mouth of Trinity River to separate at this point, as the services of an escort were no longer necessary in conferring with the Indians living around the settlements. I accordingly marched on the 24th having halted three days in order to procure a supply of provisions and give rest to the animals.

This beautiful valley, some thirty miles in length, extends along either bank of the river, and from the productive character of its soil, and its proximity to a large mining region, is fast attracting attention, and rapidly filling up with actual settlers. It was a matter of regret that no suitable means were at hand to ascertain its elevation above the sea, which must be very great to judge from the constant and rapid fall of both Scott and Klamath Rivers during their whole extent—Even at that season water froze in ordinary camp vessels every night during our stay there.

On the 20th the command crossed the dividing ridge between Scotts and Trinity Rivers, a rugged chain of great altitude and difficulty of passage, from the rocky character of its surface and being often blocked up with snow even during the early rains of winter.

The trail hence continued down Trinity River which winds through a narrow valley, and is fordable at almost any point during the dry season, though in winter a deep and rapid torrent, impassable for pack trains, as are also several creeks coming in from the west. This valley affords excellent grazing.

Such Indians who live in the mountains enclosing it are represented as hostile and thievish in character, and traders use much caution in passing through it—none were seen by our party and we had no means of determining their numbers or conditions.

On the 29th the high dividing ridge between this and Sacramento rivers was crossed at French Gulch or Dry Creek, a tributary of the latter, and the next day encamped at Shasta City, better known as Reading Spring, distant 100 miles from the foot of Scotts Valley, and the highest point in that direction accessible for loaded wagons.[109]

The route hence was along the usual road down the Sacramento valley and on the right bank of the river, the command arriving at Benicia,

[109]Reading Springs was named for Pierson B. Reading (see note 5).

196 miles from the Spring on the 9th of November. The whole circuit marched over from Sonoma, the point of departure, to the Depot at Benicia may be estimated at about 720 miles. At this season, no grazing was found from the summit of Trinity Mountain to the end of the march, which rendered it necessary to purchase forage daily during that period, increasing materially the aggregate cost of the expedition.

From my intercourse with the natives inhabiting the country adjacent to our line of march, and from as careful observation of their dispositions and habits, as circumstances would permit, I have no hesitation in saying that although thievish and suspicious in their nature, that they are more docile and more readily controlled by judicious arrangements than any tribes I have ever met with.—Mutual injuries have been sustained by this people and the miners, traders, etc. passing through their domain frequently resulting in death and creating a feeling of inveterate hostility among certain whites, which in all probability will lead to further outrages. It is to allay this feeling that I would respectfully suggest the movement of small parties of *soldiers* among these tribes, where this mutual desire for revenge is known to exist—particularly along Redwood, Trinity, & Klamath rivers.

With the tribes who agreed to treaties of peace and friendship, stipulations were made to furnish them with supplies of food for a limited period, mainly consisting of fresh beef and flour, and I deem it my duty to call attention to the loose manner in which those supplies are furnished and most urgently to recommend that a commissary or other suitable officer of [——].

[Captain Wessells no doubt signed the report, as commander of the expedition, as much of the information is in the first-person vernacular. But the last page with his signature is missing from the document.]

CHAPTER 5

Derby's Last Survey: The Lower Colorado River Valley 1850–1851

DERBY HAD JUST FINISHED HIS MAP OF THE TULARE LAKES region when he was ordered to survey the lower Colorado River valley. The army had been attempting to sustain a military presence at Fort Yuma, a crossing of the Colorado that a goodly number of emigrant trains were using. A serious problem involved supplying the fort, as most of the food and goods necessary to sustain the garrison came in from San Diego by wagon over terrible roads. General Smith hoped that shipping food via the Gulf of California might prove feasible and even be a cheaper means than using the overland route.

The journal that follows includes considerable new information on the lower Colorado valley and its Native inhabitants. Thought to be hostile, the Indians whom Derby encountered turned out to be quite friendly. That in itself was important information, given the fact that nearly everyone encountered in the Mexican communities bordering the gulf thought the opposite was true. Unfortunately, the sailing ship that the army ordered to conduct the survey also discovered problems with navigating the mouth of the river, given the strong currents, and the ship barely avoided a catastrophe. But the river was easily mastered by steamboats, the first one plying the river the very next year. Following Derby's report, steamboat traffic increased dramatically, and some half-dozen boats were working the river to Fort Yuma by the time of the Civil War. While the army probably would have discovered the feasibility of using the river to supply the fort later, Derby's report optimistically suggested that it could easily be done.

While Derby's earlier journals, two of which were never published, sparked little interest in the East, the journal from the Colorado exploration proved the opposite. Indeed, his journal was rushed east and quickly published in the serial set. The correspondence that accompanied it attests to this interest and is included below, along with the entire journal.

Letter from C. M. Conrad to William R. King

WAR DEPARTMENT[1]
Washington, June 15, 1852

Sir: In compliance with the resolution of the Senate of the 10th ultimo, I have the honor to transmit herewith a report of the colonel of topographical engineers, with a copy of the reconnoissance [*sic*] of the Gulf of California and the Colorado river, made in 1850 and 1851 by Lieutenant Derby, of the topographical engineers.

The report of the reconnoissance of Tule lake, also called for by the resolution, has not yet been received at this department.

Very respectfully, your obedient servant,
C. M. Conrad,
Secretary of War

HON. WM. R. KING
President of the Senate

---

[1]The journal that follows (and the map included in the hardback edition of this book) was originally filed in NARG 393, LR, DP. While the map was returned to the file, the journal apparently was not, and the only usable copy is found in the serial set at Senate Executive Document 81, 32nd Congress, 1st session. The journal and map were published in the now long out-of-print *Derby's Report on the Opening of the Colorado, 1850–1851*, ed. Odie B. Faulk (Albuquerque: University of New Mexico Press, 1969).

Letter from J. J. Abert to C. M. Conrad

BUREAU OF TOPOGRAPHICAL ENGINEERS
Washington, June 14, 1852

Sir: A resolution of the Senate of the 10th of May, calls for a copy "of the reconnoissance of the Gulf of California and the Colorado river made in 1850–'51, by Lieutenant Derby, of the corps of topographical engineers," together with a copy of the report made by said officer to the department in relation to the Tulare lake and its vicinity.

In obedience I have the honor to submit a copy of the reconnoissance of the Gulf of California and of the Colorado river; but it is not in my power to send a copy of the report of the reconnoissance of Tulare lake, as it has not been received at this [time], or by my inquiries at any other office in this city.

Respectfully, sir, your obedient servant,
J. J. Abert
Colonel Corps Topographical Engineers

HON. C. M. CONRAD
Secretary of War

REPORT OF THE EXPEDITION OF THE UNITED STATES TRANSport "Invincible," (Capt. A. H. Wilcox) made by order of Major General P. F. Smith, commanding Pacific division, to the Gulf of California and river Colorado, during the months of November, December, January, February and March, 1850 and 1851. By Geo. H. Derby, Brevet 1st Lieut., Topographical Engineers.

Orders from General Hooker to Lieutenant Derby

HEADQUARTERS, THIRD DIVISION
Sonoma, October 11, 1850

SIR: The schooner Invincible is to be dispatched by the chief assistant quartermaster of the division to attempt the entrance of the Colorado from the Gulf of California, and the ascent of the river as far as the post to be established near the junction of the Gila, with a view to establish that as a route of supply.

As this will afford a good opportunity of determining with accuracy many points in the geography of that part of the country of great future consequence, you will proceed to San Diego to embark in the vessel that may be sent on that service.

You will fix by observations and deductions as far as possible all of the points about the head of the Gulf of California important to the navigation of the Colorado. You will record the soundings when they may have been procured, especially at the entrance of the river and in its course upward. You will also ascertain as far as practicable the rise and fall of the tide, and the time of high water at the full and change of the moon at the mouth of the river. Likewise record the character and topography of the country at the head of the gulf and the river, as far as falls within your observation, and make a sketch of all the coast and shores in that neighborhood; and procure all such other information as may be of use to the service and the public generally.

The principal object of the expedition is to open a route of transportation by water to the mouth of the Gila, for the supply of the post to be established there. There are no means for an accurate or extensive survey, but the opportunity of gathering much useful information must not be lost, and a skillful and industrious officer is needed for that purpose. But your movements must be subordinate to those of the vessel, and her principal object must never be abandoned for the sake of even the most interesting results not connected with it.

As no military escort can be furnished, no risk will be run in those sections infested by hostile Indians.

Very respectfully, your obedient servant,
J. Hooker
Assistant Adjutant General

Lieut. G. H. Derby
Topographical Engineers, Sonoma

Derby's Survey of the Lower Colorado River Valley November 1850–February 1851

Sonoma, California
March 31, 1851

Colonel: In compliance with your instructions I left Sonoma, arriving at Benicia on the 12th of October ultimo, where I had an interview with Major Robert Allen, the chief quartermaster of the division, who supplied me with funds for any necessary expenditure that might occur during the expedition.[2]

I proceeded to San Francisco on the 13th, intending to take passage immediately for San Diego, but finding that there was no steamer to sail upon the 15th, and there being no other method of transportation available, I was compelled to remain until the arrival of the schooner Invincible, which took place on the 31st.

I had taken from the topographical office at division headquarters, one chronometer, (Bliss & Creighton, No. 1468,) two sextants, and two Schmalcalder's compasses, and was accompanied by Mr. Walter H. Ferrier, a young gentleman who had been employed to assist me in making astronomical observations, and in such surveys as might be practicable. I embraced the opportunity of the unavoidable delay at San Francisco to have my sextants thoroughly adjusted, and placed my chronometer with Captain Alexander Frazer, of the revenue service, who had established an observatory at Rincon point for rating those instruments. He discovered in a few days that it was working very irregularly, owing to some internal derangement, and I was obliged to have it removed and repaired. As I had no time to have it satisfactorily rated after its repair, and had an opportunity to procure an excellent instrument at a reasonable price, I purchased chronometer No. 1430, Parkinson and Frodsham, which has performed with the most satisfactory results during the expedition.[3]

---

[2]Major Robert Allen graduated from West Point in 1836, served with distinction in the Mexican War, and thereafter entered the Quartermaster's Corps. After his service in California, he directed supply efforts in the entire Mississippi River valley during the Civil War and ultimately retired a major general.

[3]Captain Alexander Fraser (Derby spells it Frazer) joined the United States Revenue *(continued)*

The transport Invincible, Captain A. H. Wilcox, which had been designated for the service, is a top-sail schooner of 120 tons burden, and draws when light between eight and nine feet of water. She had a crew of nine seamen, cook, steward and two mates, and was supplied with provisions for a four months' cruise. An old iron twelve-pounder carronade (captured from the Mexicans at Tabasco) and six muskets with flint locks comprised the extent of her armament.[4]

Everything being ready for our departure, we sailed from San Francisco on the 1st November with a fine breeze from the northwest, before which we rapidly ran down the coast, and passing the island of Santa Barbara upon the 2nd, and St. Clements upon the 3d, we came to anchor upon the 4th at 11 a.m., off the beach or "Plaza" of San Diego.[5] Here the vessel was detained until the 15th discharging cargo, taking in ballast and repairing her copper, during which interval I obtained a copy of Imray's chart of the Gulf of California and Lieutenant Hardy's "Travels in Mexico," which last, though very inaccurate in many important points, was nevertheless of much assistance to us in the navigation of the river.[6] I had conversation also with Colonel Williams,[7] an old well

---

Service in 1832, ultimately being assigned to California, where revenue collection was being organized as early as 1849. A sailor by training, he also designed a revolutionary cutter in the 1850s, known to history as the *Harriet Lane.* Both Parkinson and Frodsham (London) and Bliss and Creighton (New York) were known for making maritime chronometers, which were portable timepieces that were so accurate they could be used to determine longitude and latitude by means of celestial observation.

[4]Captain A. H. Wilcox was attached to the army's commissary department, which used ships to transport supplies into California. Wilcox later formed a company to run steamers on the Colorado River called the "Colorado Steam and Navigation Company." It started service in 1852 and supplied Fort Yuma. Later, steamers were transporting hides and minerals out of the upper Colorado River. A carronade was a short, smoothbore, cast-iron cannon, useful only for bombardment at close range. Rifled cannon replaced carronades in the 1840s. See Richard E. Lingenfelter, *Steamboats on the Colorado River, 1852–1916* (Tucson: University of Arizona Press, 1978).

[5]Santa Barbara Island is in the Channel Islands, off the coast of Santa Barbara, California, while Derby's "St. Clement" Island is actually San Clemente Island, off the coast of San Clemente, California, just south of Los Angeles.

[6]Robert William Hardy, *Travels in the Interior of Mexico, in 1825, 1826, 1827, and 1828* (London: Henry Coburn and Richard Bentley, 1829).

[7]This is likely Isaac, also known as Julian W. Williams, an early trapper who had a colorful life. He reached New Mexico in 1823, then joined Ewing Young on his expedition into California in 1832. It was likely during this trip that Williams explored the Colorado River. Williams married the daughter of Antonio Maria Lugo, who had a large Mexican land grant, consisting of land formerly part of San Gabriel Mission. He died at the ranch in 1856. See also Farquhar, "The Topographical Reports of Lt. George H. Derby," 45.

known resident of Los Angeles, who stated that he had explored the banks of the Colorado nearly to its entrance into the gulf, and gave me such fearful accounts of the ferocity of the Indians inhabiting the country in the vicinity of the river, that I applied to Lieutenant Colonel Magruder,[8] commanding the light battery stationed at San Diego, for some addition to our armament, which, however, he was unable to supply. Having received on board ten thousand rations for the command stationed at the junction of the Gila and Colorado, we sailed from San Diego on the afternoon of the 15th, and on the 17th passed the island of Guadaloupe, situated in latitude 29° 20'.[9]

This island is about fifteen miles in length and five in width. It is rocky and mountainous, but covered with vegetation, and is represented to be thickly inhabited by wild goats of unusual size. Water is found upon the eastern shore, and the island is frequently visited by small vessels engaged in the capture of the sea-elephant, numbers of which animals are found upon its coast. As we passed it at a distance of thirty miles, I was unable to make a sketch of its outline.[10] On the 19th we passed the Alijos rocks, bearing S.S.W. and distant about ten miles. This remarkable reef, first discovered in 1710, is about four miles in length, the highest rock being about 110 feet above the surface.[11] Its general direction is N.E. and S.W. The entire reef is so thickly covered with guano that we found it impossible to form any conjecture as to its geological character. The surf dashing over it in a high sea is a magnificent spectacle when beheld at a safe distance, but the vigias surrounding would render a close view dangerous.[12] I believe no one has ever attempted a landing. The wind continuing favorable, we made the southern extremity of the peninsula of California upon the 22nd at noon, Cape Falso bearing N. by W. and Cape St. Lucas N.E. by E. at

---

[8]John Bankhead Magruder graduated from West Point in 1830 and later served as a general for the Confederacy during the Civil War.

[9]Guadeloupe Island is off the coast of Baja California, directly west of Ensenada.

[10]Reference is here made to the northern elephant seal, which was hunted nearly to extinction by the late nineteenth century. The name comes from the proboscis on the upper head of the male, which looks like a trunk.

[11]The Alijos Rocks were a well-known collection of volcanic rocks roughly two hundred miles off the coast of southern Baja, California. The rocks actually appear on maps over a hundred years before the date suggested by Derby.

[12]A *vigia* is a hazardous rock, or shoal, that is mostly underwater and unknown.

about fifteen miles distance. The land is high, bold, and in the vicinity of the capes, mountainous. The white chalky cliffs which may be seen at a great distance render the coast easy of recognition.

We ran into the little harbor of Cape St. Lucas on the 23d, anchoring in sixteen fathoms of water, within a cable's length of the shore. The harbor is but two miles in width, protected from the southeast by a high rocky bluff, but perfectly exposed to the north and west. It is extremely dangerous anchorage, as a heavy surf is continually rolling in, and no bottom is found at a hundred fathoms, until close into the beach.[13]

The wreck of the brig Friendship was lying upon the shore, which vessel was beached here three or four months before our arrival.[14] We landed through the surf, walked up to the mission, situated in a little valley a mile from the beach, and consisting of some eight or ten adobe houses thatched with straw. The valley appeared to have been under cultivation, but the inhabitants had nearly all left for the interior to avoid the cholera then raging upon the coast. I observed here a fine herd of cattle, and many pigs, goats and fowls. The number of inhabitants I was informed was about fifty. With the exception of the little valley under cultivation, which is watered by a small stream running from the hills and emptying into the sea at this point, the country is mountainous and sterile, and covered with various specie of cacti, which grow to an immense size. Returning to the vessel I noticed upon the beach a large *copper* anchor weighing probably eighteen hundred weight; it was evidently a vestage [*sic*] of antiquity, being singularly shaped, with a remarkably long shank and narrow flukes. I was informed by an American resident of the place (Mr. Wilks) that it was fished up in the harbor three or four years ago. It was probably lost here by some of the early Spanish explorers of California. Leaving Cape St. Lucas with the evening land breeze, we passed Cape Palmo on the 24th, bearing west about thirty miles distant, and on the 25th at noon passed within ten miles of the small island of San Ignacio in

[13]Better known today as a famous vacation spot, the ship landed at Cabo San Lucas, on the tip of Baja, California.

[14]A number of brigs, or brigantines, were named *Friendship* in the early nineteenth century.

latitude 23° 39' 30".[15] The island, situated on the eastern side of the gulf, is a barron [*sic*] rock, and entirely uninhabited; it is about five miles long and two or three in width.[16]

We experienced a heavy gale from the S.W. commencing on the 26th, which gradually wore round to the N.W. and brought us off the entrance to the harbor of Guyamas [Guaymas] on the 28th.[17] We succeeded in entering the harbor on the 29th, where we anchored in four and a half fathoms of water, with the intention of remaining a few days to obtain the rates of the chronometers, and take in a supply of water and fresh provisions. I hoped also to obtain some information regarding the navigation of the gulf, from the residents of Guyamas engaged in the pearl fishery, an other persons likely to be acquainted with it.

The harbor of Guyamas is one of the best upon the coast; it is perfectly land locked and protected by its numerous islands from every quarter. It has been so often and so well surveyed, that is it unnecessary for me to enter into a minute description of its merits. It is to be regretted, however, that it has not more water, from four to five fathoms being the average depth, which is insufficient for a ship of the line, or even a first class frigate. We found Guyamas a dirty place, with a dirty population of about 1,500 or 2,000. The houses being built of adobe with the roofs sloping toward the interior, have a very unfinished appearance, and from the harbor the town presents the appearance of having been abandoned when half built. There are two small piers in a ruined condition; and near the landing a large pile of earth, surmounted by two or three crumbling walls, over which floats from a lofty staff the flag of Mexico, marks the site of the adobe fort knocked down by the guns of the Dale during the late war.[18] There are several wealthy individuals in Guyamas, who monopolize the whole of the business with the interior of Sonora, but the mass of

---

[15] The Bahía de las Palmas is on the southeastern shore of the Baja peninsula, near present-day Santiago.

[16] San Ignacio Island is the site of Laguna San Ignacio, just off the coast from the town of San Ignacio. It is known worldwide today as the laguna where gray whales come to mate.

[17] The Mexican port of Guaymas, Sonora, which Derby misspells.

[18] The *Dale* was a sloop of war, commissioned in 1839. She was on duty off Chile, protecting American commerce and whaling, when the Mexican War broke out. She landed marines at both Guaymas and Mulegé in 1847, capturing both cities.

the population are in a state of wretched poverty. One or two small vessels from San Francisco were lying in the harbor, the proprietors of which had purchased everything in the shape of fresh provisions to be obtained. Sheep in large numbers they had purchased from the interior of Sonora with the intention of landing them at Molexe, on the California coast, thence driving them overland to San Francisco.[19] Fowls, turkeys, ducks, goats, everything that could be purchased for one real and sold for twenty had disappeared, causing us no little difficulty in obtaining fresh provisions, even at comparatively high prices. The water at Guyamas is obtained from wells, and is slightly brackish. Excellent oysters are brought from the river Yaqui, which empties into the bay about twenty miles south of the town, and sold to the shipping at a dollar a bushel; the Mexicans, however, make no use of them.[20] We found the cholera raging at Guyamas on our arrival, many deaths occurring daily. It appeared to be confined to the native population entirely, which, considering their dirty habits and mode of living, was by no means to be wondered at. Dr. W. L. White, an excellent physician from San Francisco, had recently arrived, and was engaged in a very successful practice.[21] We obtained from him such medicines as were deemed necessary under the circumstances, but fortunately had no occasion to use them. I commenced a series of daily observations upon one of the small islands in the harbor, for the rates of the chronometers, which we succeeded in ascertaining with accuracy.

Immediately upon our arrival I called upon the American consul, Mr. John Robinson, who treated us with civility, and M. Cubillas, the consul of France, who received us with great attention and hospitality, and took much interest in the expedition. I also visited Mr. Thomas Spence, an old English resident, who was the agent of Lieutenant Hardy for the purchase of the "Bruja" for his expedition in 1829, and who had frequently made trips towards the head of the gulf himself, though never going far north of the island of Tiburon. He very kindly

[19]Derby had already passed the Bahía de la Paz, well to the south, which Hardy had identified as the Bay of Molexe. See Hardy, *Travels in the Interior of Mexico,* 253.

[20]The Yaqui River emptied at Guaymas, being named for the Indians of that name.

[21]Dr. White fails to appear in any other record.

gave me all the information in his power in regard to the navigation of the gulf and the small harbors upon the coast north of Guyamas.[22] All of these gentlemen united in saying that the country near the head of the gulf and bordering the river Colorado was inhabited by hostile Indians, from whom we might expect much trouble; and on becoming acquainted with the weakness of our armament, strenuously advised an increase of that and in the number of our crew. I therefore purchased from the captain of a Spanish schooner lying in the harbor two small swivels and eight carbines, with the necessary ammunition, and the captain engaged three additional seamen, increasing the number of our crew to twelve. This force I deemed sufficient for any emergency that might arise.

The time of high water at "full and change" of the moon at Guyamas is not very accurately determined; it is, however, between 8 and 9 a.m. The ordinary rise of spring tides is six feet, neap tides four feet.[23] The phenomena of four tides in twenty-four hours has repeatedly occurred here, as I am credibly informed. The prevailing winds in May, June and July are from the southeast and southwest. The thermometer during the summer months range from 92° to 98° Fahrenheit, the maximum 110°; during winter from 56° to 60°, minimum 45°. Guyamas is in latitude 27° 54' N., longitude 110° 49' 10" W.

We sailed from Guyamas on Friday, the 6th December, and for three days experienced a very heavy gale from the northeast, against which the vessel struggled under double-reefed sails without making much progress to the northward. We passed the small islands of Rolasco, situated about twenty miles west of Guyamas, on the 6th, and on the

[22]In smaller towns like Guaymas, businessmen were often selected to serve as consuls. Fernando Cubillas was heavily involved in the mining industry in Mexico. John A. Robinson was also a businessman. While the relationship is not certain, Robinson's Landing on the west bank of the mouth of the Colorado River was likely named for him. After Derby's report, entrepreneurs like Robinson almost immediately thereafter opened the Colorado to commercial traffic. While Robinson's Landing became the initial port of entry to the river where supplies for Fort Yuma were landed, other locations were also soon found. Genealogical information on the Internet suggests that Thomas Spence was from Scotland and had married a Mexican woman in Guaymas, and he too became involved in commerce. Guaymas was also important as a port where trading vessels from the Orient landed.

[23]The neap tide is a moderate tide halfway between one spring tide and the next.

7th were within ten miles of the island of Tortuga, bearing from us S. ½° W. This island is about six miles in length, and like most others in the gulf appears rocky, barren, and uninhabited by animals.[24]

The violence of the wind abated on the 8th, and on the 9th we found ourselves off the large bay represented on the chart as lying between the Cape of Trinidad and San Miguel, on the California shore. Here we observed with glasses a very extensive plain, covered with cactus. On the 10th we passed the island of "San Juan Pedro Martyr," the longitude of which we determined to be 111° 54'.[25] The island is a barren rock, about five miles long, situated nearly in the middle of the gulf. Far beyond, to the northeast, we saw the rugged outline of the celebrated island of Tiburon. This island has long been known as the abode of the Ceres Indians, a small tribe of about five hundred, who are represented as extremely hostile, and invariably opposing any attempt at landing; they are said to be armed with poisoned arrows.[26] A rich bed of pearl oysters is said to exist between this island and the coast of Sonoma, and there are accounts of rich gold mines upon the island; but as no one is ever known to have landed there, it is difficult to understand how the fact was ascertained. We should have landed and attempted an examination of the island, but the wind being contrary compelled us to lay over to the California shore, and we passed it in the night.

On the 11th we passed the island of San Estevan, which presents an exceedingly wild and sterile appearance, resembling the generality of the gulf coast on the California side. The appearance of the water between this island and that of San Lorento [Lorenzo] led us to suppose that we were on soundings, but on trying the deep sea lead we found no bottom at one hundred fathoms.[27]

---

[24]Both islands are volcanic in nature.

[25]Derby references islands that are on the eastern short of Baja. While the island of "San Juan Pedro Martyr" is mentioned, the name has likely disappeared today. Just beyond it on Baja begins the Sierra de San Pedro Mártir National Park, a mountain range that extends up the spine of Baja with peaks reaching ten thousand feet.

[26]These are the Seri Indians. See Alfred L. Kroeber, *The Seri*, Paper no. 6 (Los Angeles: Southwest Museum, 1931); De Grazia and William Neil Smith, *The Seri Indians: A Primitive People of Tiburon Island in the Gulf of California* (Flagstaff, Arizona: Northland Press, 1970); Thomas E. Sheridan, *Empire of Sand: The Seri Indians and the Struggle for Spanish Sonora, 1645–1803* (Tucson: University of Arizona Press, 1999).

[27]These islands are just to the west of the larger Tiburon Island.

On the 12th, we passed near three islands (to the south of the large island of Los Angeles), which are not put down on the chart, and of which we had no previous information.[28] They are composed of coarse clay slate, colored trap, and trap tuffa, covered with cacti. Immense quantities of seals are found upon these, as well as most of the other islands in the gulf. I named this group "Allen's Islands," as a compliment to Major Robert Allen, chief quartermaster of the Division. The longitude of the group is about 112° 40' W., latitude 28° 55' N., and soundings in the vicinity from four to seventeen fathoms.

Angelos island, which we also passed on the 12th, is separated from the California shore by Whale's channel, which is about ten or fifteen miles wide. The island is about thirty miles in length, mountainous, rocky, and uninhabited. It lies between 29° 10' and 20° 50' N., and in longitude about 113° W.[29]

We had received information in Guyamas of a small bay somewhere upon the California coast, in this vicinity, where fresh water could be obtained, and on the 13th, having rounded the northern extremity of Angeles island, we ran before the wind down Whale's channel, (where we observed a southerly current of about three miles per hour), in search of it.

We discovered a large island lying close to the California shore, off the southern extremity of Angeles island, in latitude 29° 3', longitude 112° 59' W., which, not being put down upon any chart, I named Smith's island, in compliment to the general commanding, by whose order the expedition was undertaken.[30]

This island is about a mile north of Angeles bay, which we found and entered on the 14th.[31] This is a small indentation in the coast, about four miles in width, slightly open to the eastward, with a sandy bottom, in about ten fathoms of water. The adjacent country is extremely wild and rugged. The hills are covered with huge boulders, which are coarsely aggregated masses of quartz, mica, and feldspar, and I noticed several isolated blocks of grey granite and sienite. There are

[28]Derby's Los Angeles Island is actually Isla Ángel de la Guardia.

[29]Wales Island was in between the larger Isla Ángel de la Guardia and the mainland.

[30]Reference is here made to General Persifor F. Smith, Derby's commanding officer.

[31]Better known today as Bahía de Los Angeles.

three springs of slightly brackish water at the foot of the highest hill, and a bed of most excellent oysters is exposed at low tide. The water may be found by noticing the reeds which grow about it, and which are the only green things to be seen in the vicinity.

The deep and rugged channels running down the sides of the mountains, and the immense pile of boulders forming their sides, would seem to indicate that at certain seasons of the year great quantities of water must fall here; but at the time of our visit everything appeared dry and parched. There are many varieties of cactus in the vicinity, but no nourishing herbs or grasses, and we saw no traces of game, but a few tracks of coyotes and wild goats near the spring. The bay is probably well known to the people inhabiting the interior, as we found many traces of old encampments, piles of oyster shells, heaps of ashes, and many mule-tracks leading to the southward. There are plenty of turtle in the harbor but we did not succeed in taking any. The sketch will give an idea of the character and topography of the bay, though by no means the result of an accurate survey.

We remained at Angeles bay until the 18th, when, having filled our water-casks, we weighed anchor and stood out from our anchorage into Whale's channel. It being low tide we noticed several sunken rocks at the entrance, which we had not perceived on coming in. The rise and fall at Angeles bay is about fifteen feet. Its longitude is 113° 25' 30" W., latitude 29° 5' N. We had rain on the 19th in squalls, and on the 20th a gale from the northwest, against which we steadily beat up for the head of the gulf. On the 21st we sounded when about five miles from the California shore, and found the depth from twenty to twenty-five fathoms. On the 22nd we passed within four miles of a very remarkable rock, which at a distance has the appearance of a ship with all sail set. I named it the "Ship rock," but am inclined to think it identical with Hardy's "Clarence island," although twenty miles to the westward of the position which he lays down. We sailed over the latitude and longitude of his Clarence island, but saw nothing on it.[32] The Ship rock is apparently a mile in circumference, about two hundred feet in height and perfectly white with guano.

[32]There are today nearly a hundred islands in the Sea of Cortez, and only a handful have actual names. Most of the names that Derby offers have long since disappeared.

We sounded continually on the 22nd, and found a depth of twenty fathoms, with a sandy bottom. In the afternoon we observed a column of smoke in the distance, bearing N. 60° W., which we afterwards ascertained to be caused by the Indians burning the reeds on the river Colorado. On the 23d the land was plainly discernible on both coasts of the gulf; on the California side bold and mountainous, but on the Mexican low and sandy. Sounding continually, we found about thirteen fathoms water, and we noticed constantly masses of reeds, trunks of trees, &c., floating past us, and at 6 P.M., the wind having died away, we dropped anchor in *six fathoms water.* We found by throwing the log, a current setting southward about a mile and a half per hour. The soundings from the 21st had shown that the bottom was a fine blue pipe-clay, which (mentioned by Hardy) indicated our approach to the mouth of the river. On the 24th we weighed anchor and ran over to the California shore, the water shoaling [*sic*] gradually from six fathoms to ten feet, and at night we anchored between Montague island (which we readily recognized by Hardy's description, see chart) and the main land.[33] The country here is low, flat, and covered with dwarf reed and coarse grass. Thousands of trunks of trees lie scattered over its surface as far as the eye can reach, showing that it must be entirely overflowed at the season of freshets.

We landed on the 25th, both on Montague island and the main, and found the soil clay detritus and the vegetation scanty. The two islands situated in the mouth of the river, called by Hardy Montague and Gore, are low, flat and sandy. They are separated by a channel about one mile in width, and extremely shallow. They are evidently formed by the accumulation of the sand and detritus from the river, and are gradually increasing in size. We commenced the survey of the river upon the 25th, which we continued from day to day, as we ascended. On the 27th, by taking advantage of the tide, we had succeeded in reaching Unwin's point, off which we anchored, and noticing the smoke of several fires in the distance, which we hoped might be at Major Heintzelman's encampment, fired several guns to warn him of our approach. The log gave us at this point a current at ebb tide of four

[33]At this point, most of the geographic locations mentioned by Derby, including Montague Island, which sat at the mouth of the Colorado River, are located on his map.

and a half miles per hour, which we found to be the average velocity, except at spring tides, when it is much increased.[34]

On the 28th we left our anchorage at 7 A.M., and beat up with the flood tide until about 9, when we grounded off Charles's point. The soundings had been from one and a half to three and three-fourth fathoms. At this point we found the water devoid of any brackish taste, but extremely muddy, resembling in its character that of the Mississippi river. We continued daily up the river, sounding the channel at low water, and marking it by stakes, starting with the flood-tide and floating with it until the ebb set in, when we anchored and went on shore to continue our survey. The shores of the river (here from two to four miles wide) continue of the same character, the grass growing somewhat more thickly as we ascended, however; and upon "Green Hithe point," off which we anchored upon the 1st of January, we found a thick growth of artemisia.[35] We had experienced no little difficulty in crossing from Charles's point to Green Hithe point; the channel is extremely narrow, and nearly at right angles to the direction of the river, and we had twice been swept from it by the tide and grounded upon the bank above. Upon the 1st, however, we were favored with a strong breeze, taking advantage of which we soon crossed and anchored in the channel upon the southern side. We landed upon Green Hithe point (formed by two small indentations in the crumbling clayey bank) and found the land on the south bank to be a perfectly level plain, (the soil clay detritus,) extending to the south and west for miles, and intersected by numerous sloughs, apparently filled by every spring tide. The river westward is bounded by the high hills of the lower California chain, at the distance apparently of fifteen or twenty miles. Trunks and limbs of large trees, some recently

[34]Samuel Peter Heintzelman graduated from West Point in 1826 and had a long army career, serving in the Seminole Indian War, the Mexico War, and the Civil War. He led an expedition into the Colorado River valley to stop the raids of Yuma Indians on emigrants in fall 1851 and later established Fort Yuma. The Derby expedition was expected to meet Heintzelman, hopefully not long after reaching the mouth of the Colorado. Neither Heintzelman nor Derby seemed to know that the fort was well up the river from the mouth.

[35]See Derby's map for "Greenhithe Point," which Derby spells differently than Hardy, as well as Charles Point.

deposited, others in an advanced state of decay, are thickly scattered about, evidently left by the spring freshets. The river bank is abrupt and about twenty feet in height; the water is gradually undermining it, and large pieces are continually dropping off, falling into the water with a sound as loud and not unlike the distant report of a musket. We discovered two new islands above Green Hithe point; they are low and sandy, separated by a small channel and covered with coarse grass; they have undoubtedly been formed within a few years. The bed of the river is filled with quicksand, and its current at the time of spring tides is so strong that the channel is continuously changing. I named these islands Gull and Pelican respectively, from the numerous flocks of these birds continually hovering or afloat in their vicinity. We discovered the remains of a rude Indian hut near the shore, and observed many tracks of horses in the vicinity.

During the night of the 1st the vessel grounding at ebb tide, swung around upon her heel, and thumping violently was carried by the tide (dragging the anchor) some two or three miles, grounding finally upon the shoal of Gull island; at flood-tide sail was made on her as soon as she floated, and we succeeded in getting her back into the channel. As the vessel grounded at every ebb-tide, and on the return of the water was violently swung around, thumping on her bottom and swinging on her anchor, I began to see that it would be neither prudent, or in fact possible, to ascend the river much higher, and we accordingly commenced making preparation for a boating expedition; the tides were now on the increase, and on the 2nd, for the first time, the singular phenomena of the tide coming in, in a bore or wave, while the ebb was still rushing past the vessel towards the gulf. On looking in the direction of Green Hithe point, a bank of water some four feet in height, extending clear across the river, was seen approaching us with equal velocity; this huge comber moved steadily onward, occasionally breaking as it rushed over the shoals of Gull and Pelican islands; passing the vessel, which it swung around on its course, it continued up the river. This phenomena was of daily occurrence until about the time of neap tides, and shows the truth of Hardy's assertion that "*there is no such thing as slack-water in the river Colorado.*" I took the shale-boat on the afternoon of the 2nd and proceeded up the river with the flood-tide, rounding Howard's point, (so-called by Hardy;) we found ourselves in a broad but shallow bay about four miles in width. At the northeast and northwest extremities we found the two branches of the river, the former of which we mistook for the Gila; this is in fact the main channel of the river, about a mile from its entrance, into two branches, one of which terminates in a small lagoon about four miles from its mouth, the other communicating with the river above.[36] As there is not water enough in either of these branches to float a whale-boat at low tide, it is evident that the river must have altered entirely since Lieutenant Hardy's visit, or that he never ascended it as he says he did with the Briga, a schooner of twenty-five tons. We sounded clear across the bay and found from two and a half feet to one and one-fourth fathoms. As the schooner was drawing eight feet, this settled the question as to her navigating the Colorado above this point.

On the 3rd, having ascended the river with the schooner to within four miles of the point reached the day previous with the whale-boat, we fired three heavy guns at short intervals, still hoping that Major Heintzelman's command was in hearing distance of us. The Captain, Mr. Ferrier, and myself then went on shore and walked into the country, over the level plain in the direction of the nearest range of hills. The soil is a mixture of sand and alumina, evidently overflowed and sparsely covered with coarse grass, reeds and flags, with here and there tufts of musquite [mesquite]. We observed many more foot-prints, both of men and horses, and found the remains of another rancheria; after proceeding some three miles we heard a gun from the schooner, and turning saw the ensign hoisted at the main. We immediately turned and discovered the cause of the commotion in the approach of fifteen or twenty Indians, attracted by the report of our guns in the morning. On arriving at the beach they laid aside their bows and arrows, and made signs to us to approach, and addressed us in Spanish. We took six of them including their chief on board. They were very much like all other Indians, with coarse black hair, like a horse's mane, cropped

[36]Again, the evidence is clear that Derby assumed they were close to Fort Yuma, which was located near the mouth of the Gila River, well north of them.

straight across their eyes, and with no clothing except the inevitable dirty rag worn apron fashion about their loins; they were, however, a little nastier than any Indians I had ever before seen, being beplastered from head to foot with mud, with which some of them had filled their hair; one had on a linen coat, and another, to our great joy, an infantry great coat.[37] After treating those on board with bread and molasses, I ascertained from the chief that they were acquainted with Major Heintzelman, and would carry a letter to him from me, which they said would take a day and a half; I accordingly prepared a letter, and having paid the chief in advance, sent him off with it.

We were frequently visited afterwards by these Indians and many others, to the number, perhaps of two hundred; they call themselves Co-co-pas, and live in a little village of twenty to fifty inhabitants, dispersed about near the bank of the river; the men are very tall and strongly made as a general thing, and the women are modest, well behaved, and rather good-looking; their huts are precisely like those of the California Indians, made of sticks in a spherical or oven shape, and covered with dirt.[38] They live on fish, small game, and bread made of grass seeds, and raise pumpkins, watermelons, &c., on little patches of ground which they cultivate; we found them very friendly, quiet and inoffensive; they brought fish to sell to us nearly every day, and although continually on board the vessel, we never missed even the most trifling article. Their arrows are made of reed with a pointed end of hard-wood, and their bows of willow, so that if they were disposed to be as hostile as had been falsely represented to us, they would be incapable of doing much damage. They were annoying visiters [*sic*], however, pervading every part of the vessel, gazing at every thing that was going on, begging for shirts and other articles of clothing, and afflicted with a never-to-be satisfied hunger. It was remarkable that although they had never seen a vessel before larger than their

---

[37]The use of mud was common to discourage mosquitoes.

[38]Derby's "Cocopahs" are the Cocopa people, of Yuman linguistic stock. They inhabited the west bank of the Colorado River, near its mouth where Derby found them. See Katharine Luomala, "Tipai-Ipai," in Heizer, ed., *California*, 592–609. They are also discussed in Jack Forbes, *Warriors of the Colorado: The Yumas of the Quechan Nation and Their Neighbors* (Norman: University of Oklahoma Press, 1965).

own reed rafts or balsas, they never expressed the least surprise at the "Invincible," or anything connected with her, appearing to take our arrival as a matter of every day occurrence. The villages higher up on the river own many horses, and the chiefs always rode when coming to visit us. I suppose the whole tribe may number one thousand, including men, women and children. The men frequently wear beads, rings, &c., in their noses, and paint their faces black and red with charcoal and ochre.

I sent off two other Indians with letters to Major Heintzelman, reporting our arrival, and having waited without a reply for more than a week, during which time the carpenter had been experimenting upon the long boat, putting a shaft with cranks and paddles in it, (which when tried proved of course a signal failure,) I concluded to continue the ascent of the river with a boat; the Indians who had visited us had frequently said that higher up the river lived a bad race of Indians who were thieves and murderers, (we found afterwards they meant the Yumas on the Gila;) I therefore had the long-boat armed with one of the swivels, and taking provisions for ten days, started on the 11th with the Captain, Mr. Ferrier, and six men for the junction of the Gila. We reached Arnold's point (between the main channel of the Colorado and the western slough) about 9 A.M.; here a new island one-fourth of a mile square has been formed, but at present is covered at spring-tide.[39] We rowed up the main channel (Hardy's Gila) about eleven miles, encamping at sundown on the bank.

We found the river very winding, the depth of water in the channel varied from four to fifteen feet, and the action of the tide ceases to be discernible five miles above Arnold's point, or about forty miles from the mouth of the river. The banks are lined with rushes, cane, small willows and acacia, and occasionally we observed small cotton wood or poplars; the river was full of geese and ducks, and on the banks we observed may [many] deer tracks of unusual size. The width of

[39]At this point, there is a Y in the river, the course to the east being roughly the same size as the one to the west. Hardy apparently thought it the Gila River, which was nearly a hundred miles farther north. Derby likely thought it the Gila as well, only correcting the assessment after meeting Major Heintzelman.

the river varies from two hundred yards to half a mile. We saw no Indians, although they had set fire to the rushes in our vicinity, and we noticed two of their rush balsas lying on the beach. The banks of the river were generally from ten to twenty feet high, of stratified clay. On the morning of the 12th we continued our voyage, passing about four miles above our encampment to an Indian village, the inhabitants of which to the number of one hundred, among whom I recognized some old acquaintances, sat upon the bank watching our progress in silence. We passed another rancheria in the afternoon, near which lay an old scow built of wagon bodies, which had floated down from the ferry. We encamped at sundown, having made about twelve miles. The banks we observed in many places were of fine white sand. We saw a great deal of game during the day, and we found a well-trodden horse-trail running north and south about a mile in the rear of our encampment. The velocity of the current on the river had been about one and a half and two miles per hour. Three Indians came up to our encampment from their village during the evening and presented us with a few pumpkins; they returned during the night. The banks were more thickly wooded than below, but we saw no large trees; we observed great numbers of bear and other tracks upon the banks.

We started at sunrise on the 13th, and after pulling steadily forward about five miles as we rounded a bend, I observed at a distance a white object which I took for a house, but the splashing of the water by its sides glittering in the sun soon showed it was a boat. We fired our swivel immediately, and had the pleasure of having our salute replied to by two muskets. It was Major Heintzelman, who, accompanied by Dr. Ogden and Mr. Henchelwood, (proprietors of the ferry) had left the post on the 11th, on receipt of my letters, and according to our estimate had descended the river over eighty miles.[40] As the major had made field notes of the river from the time of leaving the Gila, I considered it unnecessary for me to proceed further, and we accordingly pulled down the river in company. We reached the schooner that afternoon, and on the 14th succeeded in getting her about three miles further up

[40]The ferry over the Colorado at Yuma Crossing was begun in 1850 by L. J. F. Jaeger and George Alonso Johnson. Dr. Richard L. Ogden and Henchelwood joined them.

the river to a point two miles southeast of Arnold's point on the eastern side. Major Heintzelman left on the 15th, promising to send a party down immediately on his return to take charge of the provisions. On the night of the 16th (spring-tide) the vessel surged so violently on her chain as to part it, and we lost our smaller anchor; and on the 18th the flood-tide wave coming in with great velocity, struck the vessel or her counter as she lay aground with great violence, carrying her around with it with such force as to snap the ring of the large anchor, and we found ourselves drifting rapidly up the stream towards Arnold's point. Our situation was one of great peril, for this being the highest spring-tide, if we had grounded above, the vessel must have been lost. But the captain with admirable coolness and self-possession seized the helm, and directing her towards the bank, ordered the men to leap from the jib-boom on shore with the kedge anchor, which fortunately brought her up.

We then procured the largest logs we could find, to which, when sunk in the ground, springs were attached from the vessel, and rigged out spars to keep her at a safe distance from the bank, thus rendering her as secure as possible; our situation was, however, a very unenviable one, as, apart from any danger that might arise in our present position, we had to dread the voyage down the river, which, without anchors, and liable continually to get aground, threatened to be extremely dangerous; the accident was, moreover, to be regretted, as preventing us from touching upon the California shore in search of harbors, and at the island of Tiburon, as I had intended; on our return, every exertion was made to recover the anchors, but without success.

We remained in this position until the 27th of January, without incident, occupying the time in exploring the western slough, putting up channel marks in the river, &c. We were visited almost daily by crowds of Indians, some of them from a considerable distance up the river; they generally brought fish with them to sell, and conducted themselves in the most friendly manner. The river appears to be most plentifully stocked with fish, principally salmon, trout and suckers; a species of perch is sometimes found; all the fish have an unpleasant, muddy taste, and their flesh is soft and watery.

Lieutenant Murray arrived from Major Heintzelman's camp upon the Gila, on the 27th; he was accompanied by Dr. Ogden and a sergeant with eight men; they brought with them a wagon, and stated that the bank, for three miles above, was muddy; but as this was caused by the late overflow of the spring tides; I concluded that the ground would settle before the full moon, (15th February) and that the supplies might be safely landed; accordingly, upon the 28th, we landed all the provisions upon the bank, piling them up carefully on logs, and digging a ditch around them, and pitched tents for the men who were to remain with them. On the 29th we cast off from the bank, and floated down the river, taking advantage of the ebb tide. We got aground twice in going down, but fortunately got off at high water without accident, and on the 1st February had crossed the bar at the mouth of the river, and were fairly in the Gulf of California on our return. The bar at the mouth of the Colorado is about ten, possibly fifteen miles in width; the soundings upon it are from ten feet to four fathoms; it is a very loose, muddy bottom, and with a stiff breeze a vessel could force her way over it, even if drawing a foot or more than the lead would indicate. The distance from the junction of the Gila and Colorado to the mouth of the latter, by water, is about one hundred and four miles, owing to the many bends of the river, though the difference of latitude is but little more than half that distance. The navigation of the Gulf of California presents none of those difficulties which we had been led to anticipate. The wind we found invariably from the north-west, which, at this season of the year, is its prevailing direction; it is only during the months of June, July and August that the gales from the south-east are prevalent; except in Whale's channel we noticed none of the strong currents so frequently mentioned as existing in the gulf. It would be difficult to mistake the entrance to the river, it being in fact the head of the gulf, which gradually narrows from forty to three miles when it is joined by the river, whose turbid stream discolors the gulf for many miles to the southward, in soundings of twelve and fourteen fathoms. On the Sonora coast, however, exists an indentation some fifteen or twenty miles in depth, called by Hardy, Adair bay; the shoals of this can be seen from the mast head, a view from which would prevent one falling into error which he did of supposing it a mouth of the Colorado.[41]

The angle at the junction of the slough and the main river is called Arnold's point, and from the mouth of the river (after crossing the bar) to this point, the channel varies in depth from fifteen to thirty feet, at ordinary high tide, and may, as we have practically demonstrated, be ascended by a vessel having a draft of eight or nine feet, by taking advantage of the flood, which has a velocity of from three to five miles per hour. It is impossible to sail up, however, for although the river varies in width from three miles to six hundred yards, the channel is narrow and the navigation elsewhere obstructed by the numerous sand bars. The proper method, after passing Gore and Montague islands through the western channel of the river, is that which we adopted, to drift with the flood tide, keeping close to the highest bank, sounding continually, and anchoring before the time of high water; in this way we progressed slowly but steadily, making, perhaps, four or five miles per day, until we arrived at the point where we finally landed the stores, which I have named, as will be seen by reference to the map, "Invincible point."

Above Arnold's point the river is very circuitous, the swell of the tide rapidly decreases, the channel becomes narrow, and the water has less depth. At this season, therefore, Arnold's point may be considered the head of navigation for vessels of nine feet draught [draft]; above this point we found always from three to fifteen feet of water in the channel, whose width varied from fifty to three hundred yards; and as the river at that time was at its lowest stage, I have no hesitation in saying that it may be navigated at any season of the year by a steamboat of eighteen or twenty feet beam, drawing two and a half to three feet of water. A small stern-wheel boat, with a powerful engine and thick bottom, I would respectfully suggest to be a proper description of the vessel for this navigation, where a strong current has to be contended with, and the channel, somewhat obstructed by small snags and sawyers, is quite narrow in several places. At the present season (January, February and March) supplies from vessels arriving from the gulf may be landed near Arnold's point, upon the eastern bank, and a road being

[41]Bahía de Adair is roughly ten miles southeast of the mouth of the Colorado River.

made from the post (a work of little difficulty over a level sandy plain) they might be transported by wagons across in three days. It would be preferable, however, to establish a depot by anchoring a hulk near Charles point, laden with stores, from which a small steamboat could carry more to the post in twenty-four hours than a hundred wagons could transport in a week. Either of these methods would be far preferable to the present slow, laborious and uncertain mode of supplying wagons and pack mules across the desert from San Diego.[42]

The time of high water at "full and change" at Arnold's point is 3h. 20m. P.M., and the rise of ordinary spring tides about twelve feet; but during the season of freshets the river throughout its extent (judging from the statement of the Indians and the indications upon the banks) is at least fifteen feet higher than at the time of our visit, and the velocity of the current which, above the effect of the tide, was from one to three miles an hour, is nearly doubled; the action of the tide eases about forty miles from the mouth; the banks of the river are low, flat, and either sandy or of crumbling clay which appears to have been deposited in successive strata. Near the mouth there is no vegetation, but higher up the shores are thickly lined with cane, rushes, small willows, acacia and cotton wood, and the country in the interior covered with a coarse sharp grass. The latitude of Invincible point is 30° 50' 26", its longitude 114° 46' 43" W. The subjoined sketch, on a larger scale than the map, will give a good idea of the topography of the vicinity of Arnold's point.

The large shoal south of the point is an island, except at spring tides, and is gradually increasingly in size and height.

We reached Guyamas on our return on the 5th of February, and running into the harbor before a light breeze, let go the kedge anchor with sixty fathoms of chain, which we found would hold the vessel in still water.[43] The British frigate Meander was lying in the harbor, and noticing as we passed her that we were without anchors, her commander, Captain Keppel, R.N., very handsomely sent a boat alongside us with every offer of assistance.[44] We had, however, already accepted the kind

[42]The army followed this advice, nearly to the letter.

[43]The "kedge" anchor was a second, smaller anchor, usually stowed aft in a ship. In this case, the main anchor had been lost in the river.

[44]The HMS *Maeander* (Derby misspelled the ship's name) was captained by Sir Henry Keppel, who later became an admiral in the British navy, serving conspicuously in the Crimean War.

offer of Mr. Southworth, owner of the schooner Mary Phebe, of San Francisco, who lent us one of his anchors until we could purchase one; this we succeeded in doing at a reasonable rate, and finding every thing in the way of fresh provisions exceedingly high in consequence of the presence of the frigate, we sailed upon the 9th for San Jose upon the California coast.[45] I compared my chronometers with those of the Meander while at Guyamas; and experienced from her officers every kindness and attention in their power to offer.

We arrived at San Jose on the 11th. We found the roadstead open to the south from E.N.E. to S.W.; it is unsafe during certain seasons of the year; the bottom is very uneaven [*sic*], but the holding-ground good, being white sand. We found on our arrival the ship Glasgow, from Liverpool, and five coasters lying at anchor. The mission is situated about a mile from the beach, upon a small stream which, watering the valley of San Jose, divides into four small branches and empties into the sea at this point. The valley extends north some fifteen miles and is environed by the high and rugged mountains of the lower California range. The population of San Jose is between five and six hundred. A large number of Americans, I was informed, had settled in the valley above. The soil appears sandy, which is owing to a severe flood that occurred about eighteen months ago, washing away all the upper and richer surface into the sea, and forming the four branches to the stream which we noticed. Grapes were previously abundant; now there are but two or three vineyards. The sugar-cane and banana are cultivated here as well as sweet potatoes, pumpkins, Indian corn, &c., but the crop this season will be small, the inhabitants having abandoned their gardens during the working season and gone to the interior to avoid the cholera. Cattle, mules, swine and goats are bred in the vicinity, but of the first many perish on account of the drought which had been unusual, and was severely felt in the interior. I noticed here great numbers of birds called the secretary.[46] We sailed from San

[45]The schooner *Mary Phebe* often made port at San Francisco, sailing the spring before—in 1851—from Guaymas to San Francisco with a cargo of mainly sheep, other livestock, and "35,000" oysters. See Maritime Arrivals, San Francisco, published in *Alta California*.

[46]Just what bird species Derby is referring to is difficult to determine. The secretarybird is a famous bird of prey found in South Africa, looking somewhat like an eagle, often standing four feet tall. It is possible that Derby is actually referring to a California condor, which would have been about the same size.

Jose on the 12th, having procured some vegetables of which we stood much in need, and after a long voyage of twenty-one days without any incident worth of recording, arrived at San Francisco on the 16th day of March.[47]

I have the honor to transmit herewith an extract of my journal kept during the voyage, showing the range of the thermometer, state of the weather, direction of the wind, and geographical position of the vessel for each day during the expedition. Also, a chart of the gulf and the river Colorado, from its mouth to the junction of the Gila. The river is platted from my own notes as far north as Heintzelman's Point, above this from notes made from Major Heintzelman. The latitudes given are calculated from meridian altitudes of the sun, taken by Captain Wilcox, Mr. Ferrier and myself; the longitudes from a daily mean of nine observations taken for the error of our chronometers near 9 A.M., and when possible, as in port, from equal altitudes of the sun in the forenoon and afternoon.

I am, sir, with high respect, your obedient servant,
GEO. H. DERBY
Brevet 1st Lieut. Topographical Engineers

Lieut. Col. J. Hooker,
Adjutant General Pacific Division

[47] If the 12th is accurate, the sailing time of twenty-one days would have put them at San Francisco on March 3.

Bibliography

Government Documents

House Executive Document 17, 31st Congress, 1st session.

National Archives Record Group 75, Letters Received, California Superintendency.

National Archives Record Group 393, Letters Received, Department of the Pacific.

Senate Executive Document 47, 31st Congress, 1st session, part 2.

Senate Executive Document 61, 32nd Congress, 1st session.

Senate Executive Document 81, 32nd Congress, 1st session.

Senate Executive Document 82, 32nd Congress, 1st session.

Senate Executive Document 110, 32nd Congress, 1st session.

Senate Miscellaneous Document 148, 30th Congress, 1st session.

Books and Articles

Amero, Richard W. *Eviction at Cupa.* http://balboaparkhistory.net/glimpses.

Anderson, Gary Clayton. *Ethnic Cleansing and the Indian: The Crime That Haunts America.* Norman: University of Oklahoma Press, 2014.

Anderson, M. Kate. *Tending the Wild: Native American Knowledge and the Management of California's Natural Resources.* Berkeley: University of California Press, 2005.

Baldwin, Bruce, Douglas H. Goldman, David J. Kell, Robert Patterson, and Thomas J. Rosatti, eds. *The Jepson Manual: Vascular Plants of California.* 2nd ed. Berkeley: University of California Press, 2012.

Beebe, Rose Marie, and Robert M. Senkewicz. *Testimonios: Early California Through the Eyes of Women, 1815–1848.* Berkeley: Heyday Books, 2006.

Bryant, E. *What I Saw in California.* 1846. Reprint, Minneapolis: Ross and Haines, 1967.

Burlin, Paul T. "Peter Allen Brinsmade and the Tragic Pursuit of Pious Capitalism." In *Imperial Maine and Hawai'i: Interpretive Essays in the History of Nineteenth-Century American Expansion.* Lanham, Md.: Rowan & Littlefield, 2008. 21–56.

Carle, David. *Introduction to Water in California.* Berkeley: University of California Press, 2004.

Carr, J. *Pioneer Days in California.* 1891. Reprint, Fresno, Calif.: Mid-Cal Books, 1982.

Carson, James H. *Bright Gem of the Western Seas: California, 1846–1852.* Ed. Peter Browning. Lafayette, Calif.: Great West Books, 1991. (This is a reprint of Carson's *Recollections.*)

———. *Recollections of the California Mines: An Account of the Early Discoveries of Gold, with Anecdotes and Sketches of California and Miners' Life and a Description of the Great Tulare Valley.* Stockton, 1853.

Castro, Doris Shaw. *James H. Carson's California, 1847–1853.* New York: Viking Press, 1997.

Caughey, John W., ed. *The Emigrants' Guide to California,* by Joseph E. Ware. St. Louis: J. Halstall, 1849; reprint, Princeton: Princeton University Press, 1932.

Ciruzzi, Canice G. "Phoenix Revisited: Another Look at George Horatio Derby." *San Diego History* 26 (Spring 1980): 1–12.

Clark, William B. *Gold Districts of California.* Bulletin no. 193. Sacramento, Calif.: California Division of Mines and Geology, 1980.

Cook, Sherburne. *The Conflict between the California Indian and White Civilization.* Book 2. Berkeley: University of California Press, 1943.

———. *The Population of the California Indians, 1769–1970.* Berkeley: University of California Press, 1976.

Cowan, Robert. *Ranchos of California: A List of Spanish Concessions, 1775–1822, and Mexican Grants, 1822–1846.* Fresno, Calif.: Library Guild, 1956.

De Grazia, Ted, and William Neil Smith. *The Seri Indians: A Primitive People of Tiburón Island in the Gulf of California.* Flagstaff, Ariz.: Northland Press, 1970.

Driscoll, John K. *Rogue: A Biography of Civil War General Justus McKinstry.* Jefferson, N.C.: McFarland, 2006.

Durham, David L. *California's Geographical Names: A Gazetteer of Historical and Modern Names of the State.* Clovis, Calif.: Word Dance Press, 1998.

Emory, W. H. *Notes of a Military Reconnoissance from Fort Leavenworth, in*

Missouri, to San Diego, in California, Including Parts of Arkansas, Del Norte, and Gila Rivers. Washington, D.C.: Wendell and Van Benthuysen, 1848.

Engelhardt, Zephyrin. *San Juan Capistrano Mission.* San Francisco: Standard Printing, 1922.

Engstrand, Iris Wilson. *William Wolfskill, 1798–1866: Frontier Trapper to California Ranchero.* Glendale, Calif.: Arthur H. Clark, 1965.

Farquhar, Francis P. "The Topographical Reports of Lt. George H. Derby." *Quarterly of the California Historical Society* 11 (Summer, Spring, Fall 1932).

Faulk, Odie B., ed. *Derby's Report on the Opening of the Colorado, 1850–1851.* Albuquerque: University of New Mexico Press, 1969.

Forbes, Jack. *Warriors of the Colorado: The Yuma of the Quechan Nation and Their Neighbors.* Norman: University of Oklahoma Press, 1965.

Frazer, Robert W., ed. "Lovell's Report on the Cahuilla Indians, 1854." *San Diego Historical Quarterly* 22 (Winter 1976).

Frémont, John Charles. *Geographical Memoir upon Upper California, in Illustration of His Map of Oregon and California, Addressed to the Senate of the United States.* Washington, D.C.: Wendell and Van Benthuysen, 1848.

Gudde, Erwin G., and William Bright. *California Place Names: The Origins and Etymology of Current Geographical Names.* 4th ed. Berkeley: University of California Press, 2004.

Gudde, Erwin G., and Elisabeth K. Gudde, eds. *California Gold Camps: A Geographical and Historical Dictionary of Camps, Towns, and Localities Where Gold Was Found and Mined, Wayside Stations and Trading Centers.* Berkeley: University of California Press, 1975.

Gutiérrez, Ramon A., and Richard Orsi, eds. *Contested Eden: California Before the Gold Rush.* Berkeley: University of California Press, 1998.

Haas, Lisbeth, ed. *Pablo Tac, Indigenous Scholar: Writing on Luiseño Language and Colonial History, c. 1840.* Berkeley: University of California Press, 2011.

Hammond, George P. *The Weber Era in Stockton History.* Berkeley: University of California Press, 1982.

Hardy, Robert William. *Travels in the Interior of Mexico, in 1825, 1826, 1827, and 1828.* London: Henry Coburn and Richard Bentley, 1829.

Harlow, Neal, ed. *City of Angels and the City of the Saints.* San Marino, Calif.: Huntington Library, 1978.

Heizer, Robert, ed. *California.* Vol. 8 of *Handbook of North American Indians,* series ed. William C. Sturtevant. Washington, D.C.: Smithsonian Institution, 1978.

———. *Collected Documents on the Causes and Events in the Bloody Island Massacre of 1850.* Berkeley: University of California Archaeological Research Facility, 1973.

Hewes, Minna, and Gordon Winant Hewes, eds. *Indian Life and Customs at Mission San Luis Rey*. 1958. Reprint, Whitefish, Mont.: Literary Licensing, 2012.

Hillman, Raymond W., and Leonard A. Covello. *Cities and Towns of San Joaquin County Since 1847*. Fresno, Calif.: Panorama West Books, 1985.

History of Yuba County, California. http://yubaroots.com.

Hoffman, Ogden. *Reports of Land Cases Determined in the United States District Court of the Northern District of California*. San Francisco: Numa Hubert, 1862.

Hoover, Mildred Brooke, Hero E. Rensch, and Ethel Grace Rensch. *Historic Spots in California*. Ed. William N. Abeloe. 3rd ed. Stanford: Stanford University Press, 1966.

Hurtado, Albert L. *Indian Survival on the California Frontier*. New Haven: Yale University Press, 1988.

———. *John Sutter: A Life on the North American Frontier*. Norman: University of Oklahoma Press, 2006.

Keyes, Erasmus Darwin. *Fifty Years' Observations of Men and Events, Civil and Military*. New York: C. Scribner's Sons, 1884.

———. *Fighting Indians in Washington Territory*. 1884. Reprint, Fairfield, Wash.: Ye Galleon Press, 1988.

———. *The Rear Guard at Malvern Hill*. In R. U. Johnson and C. C. Buel, eds., *Battles and Leaders of the Civil War*. New York: Century War Series, 1884–87.

Kroeber, Alfred L. *California Place Names of Indian Origin*. Berkeley: University of California Press, 1916.

———. *Handbook of the Indians of California*. 1925. Reprint, New York: Dover, 1976.

———. *The Seri*. Paper no. 6. Los Angeles: Southwest Museum, 1931.

Larson, Elizabeth. "Bloody Island Atrocity Remembered at Saturday Ceremony." *Lake Country News* (Lakeport, Calif.), May 13, 2007.

Lingenfelter, Richard E. *Steamboats on the Colorado River, 1852–1916*. Tucson: University of Arizona Press, 1978.

Lyon, Nathaniel. *The Last Political Writings of Nathaniel Lyon, USA; with a Sketch of His Life and Service*. New York: Rudd & Carleton, 1861.

Madley, Benjamin. "California's Yuki Indians: Defining Genocide in Native American History." *Western Historical Quarterly* 34 (Autumn 2008): 303–34.

Munro-Fraser, J. D. *History of Santa Clara County, California: Including Its Geography, Topography, Climatology, etc.* San Francisco: Alleg, Bowen, 1881.

Nabokov, Peter. *Where the Lightning Strikes: The Lives of American Indian Sacred Places*. New York: Penguin, 2006.

Owens, Kenneth N. *Gold Rush Saints: California Mormons and the Great Rush for Riches*. Spokane, Wash.: Arthur H. Clark, 2004.

Philips, Christopher. *Damned Yankee: The Life of Nathaniel Lyon.* Columbia: University of Missouri Press, 1990.

Phillips, George H. *The Enduring Struggle: Indians in California History.* San Francisco: Boyd & Fraser, 1981.

———. *Indians and Indian Agents: The Origins of the Reservation System in California, 1849–1852.* Norman: University of Oklahoma Press, 1997.

Phoenix, John (pseud. of George H. Derby). *Phoenixiana; or, Sketches and Burlesques.* New York: D. Appleton, 1856.

Prucha, Francis Paul. *American Indian Policy in the Formative Years: The Indian Trade and Intercourse Acts, 1790–1834.* Lincoln: University of Nebraska Press, 1962.

Rarick, Ethan. *Desperate Passage: The Donner Party's Perilous Journey West.* New York: Oxford University Press, 2008.

Rawls, James J., and Richard J. Orsi, eds. *A Golden State: Mining and Economic Development in Gold Rush California.* Berkeley: University of California Press, 1998.

Reports of Cases Determined in the Supreme Court of the State of California. Vol. 3. San Francisco: Bancroft-Whitney, 1888.

Richman, Irving Berdine. *California Under Spain and Mexico, 1835–1847.* New York: Cooper Square Publishers, 1965.

Rolle, Andrew. *California: A History.* Wheeling, Ill.: Harlan Davidson, 2003.

Salomon, Carlos Manuel. *Pío Pico: The Last Governor of Mexican California.* Norman: University of Oklahoma Press, 2010.

Sheridan, Thomas E. *Empire of Sand: The Seri Indians and the Struggle for Spanish Sonora, 1645–1803.* Tucson: University of Arizona Press, 1999.

Shumate, Albert. *Francisco Pacheco of Pacheco Pass.* Stockton: Holt-Atherton Pacifica Center for Western Studies, 1977.

Shumway, Burgess M. *California Ranchos: Patented Private Land Grants Listed by County.* San Bernardino, Calif.: Borgo Press, 1988.

Starr, Kevin. *California: A History.* New York: Modern Library, 2005.

Stephenson, Terry E. *Don Bernardo Yorba.* Los Angeles: Glen Dawson, 1963.

Stewart, George R. *John Phoenix, Esq., the Veritable Squibob.* New York: Henry Holt, 1937.

Tinkham, George Henry. *History of Stanislaus County California: With Biographical Sketches of the Leading Men and Women of the County Who Have Been Identified with Its Growth, etc.* Los Angeles, Calif.: Historic Record, 1921.

Tuttle, Charles A. *Report of Cases Determined in the Supreme Court of the State of California.* Vol. 25. Sacramento: O. M. Clayes, 1865.

Twain, Mark (Samuel Clemens). *Roughing It.* Hartford, Conn.: American Publishing, 1872.

Tyson, Philip. *Geology and Industrial Resources of California.* Baltimore: Wm. Minifie, 1851.

Warner, Barbara R. *The Men of the Bear Flag Revolt and Their Heritage.* Glendale, Calif.: Arthur H. Clark, 1996.

Wilbur, Marguerite Eyer, ed. *The Adventures of Heinrich Lienhard.* Los Angeles: Calafia Society, 1941.

Wilson, James Grant, and John Fiske, eds. *Appletons' Cyclopaedia of American Biography.* 6 vols. New York: D. Appleton, 1888.

Wood, Ellen Lamont. *George Yount: The Kindly Host of Caymus Ranch.* San Francisco: Grabhorn Press, 1941.

Woods, Daniel S. *Sixteen Months at the Gold Diggings.* New York: Hayes & Brothers, 1851.

Index